T0137088

Network Science In Education

Catherine B. Cramer • Mason A. Porter
Hiroki Sayama • Lori Sheetz
Stephen Miles Uzzo
Editors

Network Science In Education

Transformational Approaches in Teaching and Learning

Editors
Catherine B. Cramer
Data Science Institute
Columbia University
New York, NY, USA

Mason A. Porter
Department of Mathematics
University of California Los Angeles
Los Angeles, CA, USA

Hiroki Sayama
Department of Systems Science and
Industrial Engineering
Binghamton University
State University of New York
Binghamton, NY, USA

Lori Sheetz
Center for Leadership and Diversity
in STEM
Department of Mathematical Sciences
United States Military Academy
West Point, NY, USA

Stephen Miles Uzzo
New York Hall of Science
Corona, NY, USA

ISBN 978-3-030-08407-3 ISBN 978-3-319-77237-0 (eBook)
https://doi.org/10.1007/978-3-319-77237-0

Printed on acid-free paper

This Springer imprint is published by the registered company Springer International Publishing AG
part of Springer Nature.
The registered company address is: Gewerbestrasse 11, 6330 Cham, Switzerland

Preface

A new era is dawning for learning and education. It has emerged from the urgent need to provide opportunities for lifelong learners to develop skills and habits of mind that are relevant to today's complex and interconnected society. In this new era, we recognize that extant atomistic and disconnected ways of teaching and learning do not allow new standards to be implemented effectively, nor do they help people grapple with the growing complexity of science, technology, engineering, and mathematics (STEM). *Network Science In Education: Transformational Approaches in Teaching and Learning* represents a different, interconnected way of thinking about learning and a pathway into leveraging a network paradigm for new approaches to developing methods, curricular materials, and resources for learning and teaching. It also suggests ways to gain insights into the structure of the connected nature of teaching and learning communities and curricula.

The concept of *networks*—discrete structures that consist of nodes (also called vertices, entities, actors, items, etc.) and links (also called edges, relationships, ties, connections, etc.) that connect nodes to each other—has proliferated rapidly as a way of improving the understanding of nearly every system that affects life on Earth. The systems all around us, and even inside us, often include network structures. Examples of such systems include the Internet, social media, financial systems, transportation systems, ecosystems, organizations and corporations of all kinds, friendships, kinships, schools, classrooms, learning materials, brains, immune systems, genes/proteins within a single cell, and much more. Network science—the science of connectivity—is an interdisciplinary field of research on connected systems that offers a powerful approach for conceptualizing, developing, analyzing, and understanding solutions to complex social, health, technological, and environmental problems. It is also a promising pedagogical approach, and it is becoming increasingly prominent in education practice.

Since 2004, we have facilitated the growth of this work through organizing a variety of outreach programs, conference talks, curricula and other kinds of resources, and, most recently, annual symposia on Network Science in Education for the International School and Conference on Network Science (NetSci). *Network Science In Education* explores the variety of ways that networks are being brought to learners of all ages, including: new courses, new curricula, new tools, new techniques, and even the structure of systems of education and how they work. For educators, *Network Science In Education* is intended (1) to promote network thinking among students, teachers, administrators, curriculum developers, and the general public and (2) to improve their understanding of the complex systems around us through networks. For researchers, *Network Science In Education* is intended to increase awareness both of the value of network science in learning and education and of the need to cultivate a generation of network-literate people to bring this knowledge to everyday life, leverage network science to improve discovery, and better understand our relationship with nature and with each other.

Network Science In Education represents a rapidly growing community of network science researchers and educators from around the world who have come together because of a shared passion for making network science tools and ideas accessible to everyone, everywhere. The purpose of this volume is to help expand the dialog among research and educational communities to share knowledge and realize the value of network science in a multitude of learning settings, and for all learners. The structure of this book reflects the diverse audiences for network science concepts, tools, and resources: undergraduate and graduate students, K–12 learners and teachers, and the general public (e.g., through informal learning opportunities). We also envision that this volume will contribute to helping transform the learning, teaching, and understanding of the structure and function of learning communities.

We realize that network thinking is, for many, a new kind of lens on the world, and while this "science of connectivity" is intuitive and familiar to readers in many ways, it may be helpful to gain more depth of understanding on how networks appear in many areas of human culture and in our understanding of nature. There are myriad resources available that are accessible through keyword searches on search engines. A few useful and accessible ones include:

- The Network Science in Education Resource Page (https://sites.google.com/a/binghamton.edu/netscied/teaching-learning/resources)
- The Wikipedia page for Network Science (https://en.wikipedia.org/wiki/Network_science)
- *Network Science* by Albert-László Barabási (http://barabasi.com/networksciencebook/)

We would like to acknowledge Albert-László Barabási for encouraging this work; the National Science Foundation for its ongoing interest and support; and the entire network science community for their deep interest in and concern for educating future generations.

New York, NY, USA	Catherine B. Cramer
Los Angeles, CA, USA	Mason A. Porter
Binghamton, NY, USA	Hiroki Sayama
West Point, NY, USA	Lori Sheetz
Corona, NY, USA	Stephen Miles Uzzo

Contents

Part I   Creating New Courses

An Undergraduate Mathematics Course on Networks   3
Mason A. Porter

Leading Edge Learning in Network Science   23
Ralucca Gera

Advances in Nontechnical Network Literacy: Lessons Learned
in Tertiary Education   45
Paul van der Cingel

Part II   Creating New Degree Programs

Network Science Undergraduate Minor: Building a Foundation   59
Chris Arney

Evaluation of the First US PhD Program in Network Science:
Developing Twenty-First-Century Thinkers to Meet the Challenges
of a Globalized Society   71
Evelyn Panagakou, Mark Giannini, David Lazer, Alessandro Vespignani,
and Kathryn Coronges

Europe's First PhD Program in Network Science   87
János Kertész and Balázs Vedres

Part III   Education Network Analysis

Mapping the Curricular Structure and Contents of Network
Science Courses 101
Hiroki Sayama

Pay, Position, and Partnership: Exploring Capital Resources Among a School District Leadership Team 117
Alan J. Daly, Yi-Hwa Liou, and Peter Bjorklund Jr

Part IV   Tools and Techniques

Secondary Student Mentorship and Research in Complex Networks: Process and Effects.................................... 141
Catherine B. Cramer and Lori Sheetz

The Imaginary Board of Directors Exercise: A Resource for Introducing Social Capital Theory and Practice 159
Brooke Foucault Welles

Network Visualization Literacy: Novel Approaches to Measurement and Instruction 169
Angela Zoss, Adam Maltese, Stephen Miles Uzzo, and Katy Börner

Network Science in Your Pocket 189
Toshihiro Tanizawa

Index.. 201

About the Editors

Catherine B. Cramer works at the intersection of data-driven science, learning and workforce development, specifically as it pertains to the understanding of complexity. She is currently managing industry engagement at the Columbia University Data Science Institute, working with faculty and students with a focus on applying data-driven interdisciplinary research to society's most complex problems through innovation and collaboration with industry and government.

Mason A. Porter is a Professor in the Department of Mathematics at UCLA. His research interests include a plethora of topics in complex systems, networks, nonlinear systems, and their applications. He is a Fellow of both the American Mathematical Society and the American Physical Society.

Hiroki Sayama is a Professor of Systems Science and Industrial Engineering and the Director of the Center for Collective Dynamics of Complex Systems at Binghamton University, State University of New York. His research interests include complex systems, dynamical networks, human and social dynamics, artificial life/chemistry, interactive systems, and other computer/information science related topics.

Lori Sheetz is the Director of the United States Military Academy's Center for Leadership and Diversity in STEM and a collaborator with the Network Science Center at West Point. Her research interests include effective STEM education and outreach within diverse populations traditionally underrepresented in STEM fields, bridging the STEM skills gap between standards-based curriculum and projected workplace skills, and introducing a network approach to teaching and learning to precollege students and teachers.

Stephen Miles Uzzo is the Chief Scientist for the New York Hall of Science, where he does research and development of public programs and experiences on complex science and instructional development for preservice teacher education. His background includes teaching and learning in data-driven science, computer graphics systems, engineering, and environmental science.

Part I
Creating New Courses

An Undergraduate Mathematics Course on Networks

Mason A. Porter

1 Introduction

The study of networks incorporates tools from a diverse collection of areas—such as graph theory (of course), computational linear algebra, dynamical systems, optimization, statistical physics, probability, statistics, and more—and it is important for applications in just about any area that one can imagine [1, 2]. It is thus important to teach courses on networks in mathematics, statistics, computer science, social and organizational sciences, and other disciplines. Graph theory is an old subject, and mathematics departments have taught courses in it for decades. One can also find courses on various aspects of networks in departments such as statistics, computer science, sociology, and others. Many of them have existed for quite a while, but the notion of studying the mathematics of networks—as involving subjects like graph theory, but distinct from it in crucial ways—is relatively new, and both undergraduate and graduate mathematics curricula need to include courses with such a focus.

The importance of teaching courses on networks in mathematics departments goes far beyond the establishment of the topic of "networks" as having a distinct identity from subjects such as graph theory. The study of discrete data has undergone a revolution, and people with mathematics degrees need to be well-versed in it. Many mathematics majors go on to careers in some form of data science (in academia, industry, government, and elsewhere) [3], and mathematics curricula need to prepare them for these careers. One way to do this is to offer a suite of courses to develop a "discrete structures and data science" track through degree programs in the mathematical sciences, including through a mathematics major itself.

M. A. Porter (✉)
University of California Los Angeles, Los Angeles, CA, USA

© Springer International Publishing AG, part of Springer Nature 2018
C. B. Cramer et al. (eds.), *Network Science In Education*,
https://doi.org/10.1007/978-3-319-77237-0_1

Students who undertake such a track focus predominantly on discrete structures, and they should master elements of theoretical ("pure") mathematics, statistics, applied mathematics (including mathematical modeling), computer science, and data analysis. In addition to networks (the science of connectivity), such students should learn about subjects such as optimization, probability theory, machine learning, information theory, and complex systems.

In both teaching and research, my approach to the study of networks takes the perspective of "physical applied mathematics" [2]—focusing on modeling, with an origin and practice associated most traditionally with differential-equation models of physical phenomena—and I developed my undergraduate networks course with this philosophy in mind. I put a strong emphasis on mechanistic modeling, which contrasts both with the approaches to studying networks in courses on graph theory and with those in courses in statistics and computer-science departments. My blog associated with the University of Oxford version of my networks course [4] includes links to review articles and other online sources to supplement the lecture notes and main text.

In addition to my course, numerous other existing networks courses (with the number expanding rapidly), at multiple curricular levels, are taught in a variety of departments (e.g., statistics, computer science, physics, and so on) and emphasize different topics and approaches. For some examples, see [5–18]. See Chapter 7 for a comparison of the topics and organization in many existing courses on networks [19].

The rest of this chapter is organized as follows. In Sect. 2, I overview the topics that I cover in my networks course. In Sect. 3, I discuss how my course evolved from an informal set of lectures to a masters-level course and finally to a course for both undergraduates and masters students (including a version that is only for undergraduates). I highlight a few of the challenges in teaching my course in Sect. 4, and I conclude in Sect. 5.

2   Topics

The goal of (all versions of) my course, which I first taught in the University of Oxford's mathematics department, called the "Mathematical Institute" (MI), is to survey the study of networks from the perspective of mathematical modeling and to allow students to jump into the research literature. For example, my course's learning outcomes in the 2015 blurb in the MI's undergraduate handbook read as follows:

> Students will have developed a sound knowledge and appreciation of some of the tools, concepts, and computations used in the study of networks. The study of networks is predominantly a modern subject, so the students will also be expected to develop the ability to read and understand current (2015) research papers in the field.

In Table 1, I overview the topics that I cover in my networks course, which at University of Oxford included 16 hours of lectures and in later years—after being converted to a course that is intended primarily for undergraduates—also

Table 1 An overview of the topics in my undergraduate networks course at University of Oxford. There are 16 lectures of about 50 minutes (or so) each. I covered some of the listed topics in detail. I touched upon others briefly as generalizations of ideas, concepts, models, or methods that I discussed in detail.

Unit	Examples of topics
1. Introduction and basic concepts (1–2 lectures)	Nodes, edges, adjacencies, weighted networks, unweighted networks, degree and strength, degree distributions, other types of networks
2. Small worlds (2 lectures)	Clustering coefficients, paths and geodesic paths, Watts–Strogatz networks (focus is on modeling and heuristic calculations)
3. Toy models of network formation (2 lectures)	Preferential attachment, generalizations of preferential attachment, network optimization
4. Additional summary statistics and other useful concepts (2 lectures)	Modularity and assortativity, degree–degree correlations, centrality measures, communicability, reciprocity and structural balance
5. Random graphs (2 lectures)	Erdős–Rényi graphs, configuration model, random graphs with clustering, other models of random graphs or hypergraphs, application of generating-function methods (focus is on modeling and heuristic calculations; material in this section forms an important basis for units 6 and 7)
6. Community structure and other mesoscale structures (2 lectures)	Linkage clustering, optimization of modularity and other quality functions, overlapping communities, other methods and generalizations
7. Dynamics on networks (3–4 lectures)	General ideas, models of biological and social contagions, percolation, voter and opinion models, other topics
8. Additional topics (0–2 lectures)	Examples of possibilities: games on networks, exponential random graphs, network inference, temporal networks, multilayer networks, other topics of special interest to students (depending on how much time there is and the interests of current students)

incorporated "classes" (i.e., recitation sections) to discuss problem sheets. When I am teaching, I often have a tendency to include too much information.[1] An alternative design for an introductory networks course would be to cover fewer topics but to study them in greater depth.

For most topics, I based my presentation largely on discussions in Mark Newman's textbook [1], although I drastically changed both presentation order and the relative emphases on topics. For more advanced topics, such as community structure and dynamical processes on networks, my course departed rather substantially from (and/or built substantially on) the discussions in [1]. For these capstone topics, I drew a lot of the material from survey, tutorial, and review articles [4, 20, 21]. I also extracted material from particularly instructive research articles (e.g., [22]), and I referred students to additional resources on my course blog [4]. The topics that I discussed in units (6) and (7) of the course (see Table 1) have varied over the years,

---

[1] I advocate a philosophy that students should "drink from the firehose of knowledge" (to quote a saying that I learned as an undergraduate at Caltech).

and I put some of them in homework problems only rather than in the lectures. For all units, I also discussed (at least briefly) some generalizations of ideas that are explored in [1]. For these generalizations, students (if they desire it) need to examine other sources to learn more details. In practice, covering topics (6) and (7) in a reasonable way, even at an introductory level, takes so much time that I have always chosen to spend more time on them than what is indicated in Table 1, rather than having lectures dedicated to topics from unit (8).[2]

3 Evolution of My Course

From the beginning, it was my intention to ultimately offer my networks course to undergraduates in the mathematical sciences at University of Oxford, but it started out as an informal set of lectures, which were attended by some masters students, doctoral students, and others.

3.1 Stage 1: An Informal Set of Lectures

The prehistory of my course dates to 2010. Invited by David Cai, in July 2010, I gave a set of ten lectures (of about 2.5 hours each) on "Network Dynamics" (although I covered both structure and dynamics) at an applied-mathematics summer school for masters students at Shanghai Jiao Tong University. I started adapting material from [1] and organized the material mostly as presented in Table 1 (though I added new topics to the possibilities in unit (8) as network science advanced). In practice, however, I spent way too much time on early units and ended up focusing mostly on units (1)–(4), with only a little bit of material from units (5)–(7). Using the organization that I developed for the summer school as a template, I signed a book deal to write an undergraduate textbook (which I still haven't finished) for mathematicians and other quantitative scientists, where my choice of 8 units specifically matched the 16 one-hour (technically, 50 minutes or so) lectures in a standard MI course.

At University of Oxford, I first gave my networks course as an informal set of lectures in the spring term ("Trinity term") in 2012. I taught one day a week, using a two-hour slot with a roughly 10-minute break at some natural point in the middle.[3]

---

[2] I typically mention temporal networks and multilayer networks very briefly in passing, in part because of their prominence and in part because I spend a lot of time thinking about them in my research. Additionally, unit (6) interfaces with topics like network inference, which I mention only in passing when teaching my course.

[3] At University of Oxford, it is more common to meet twice a week for "one hour" (which encompasses 50 minutes of lecturing), but my course met for one double-slot each week in most of the years that I taught it at Oxford, as I felt that this choice fit better with the 8-unit organization in Table 1.

I taught my course in the MI, and it was attended by students from the MI's Master of Science (MSc) program in Mathematical Modelling and Scientific Computation (MMSC), various doctoral students, occasional faculty members, and others. The MMSC students could use my course as a "special topic" if they wrote an extended essay on a subject that was agreed by them and me (as is standard for options courses in their program).[4] I did not assign any homework, though I pointed students to topics that they might be interested in pursuing in more detail.

3.2 Stage 2: A More Formal, Masters-Level Course

In Spring 2013, students from both the MMSC program and the MSc program in Mathematical Foundations of Computer Science (MFoCS) could take my networks course.[5] Over the years, many MFoCS students were becoming increasingly interested in applied topics, and it was desirable for my course on networks to be available for them to take as an option. To accommodate requirements for the MFoCS program, I needed to add two things: (1) homework problems for those students that went beyond what I assigned to undergraduate students (to ensure that the course was an MSc-level course) and (2) final "miniprojects" to determine student grades.

In 2013, I did not assign any homework assignment for most students, so I added a couple (three in the first year, but two in subsequent years) of homework assignments that required summarizing a research article and "refereeing" it. These assignments compel students to read papers in depth, learn how to evaluate papers and hopefully also some lessons about how to write papers, and learn good scientific citizenship (through volunteer work as referees).[6] I also met with the students to discuss each paper. The students typically did a very good job at the refereeing assignments, and paper authors to whom I showed these reports (with the students' permission) mentioned that my students' feedback was typically much more helpful than the actual referee reports that they received.

Following MFoCS rules, the students had 3 weeks at the end of a term to do their miniprojects, which are supposed to take 3–4 days of dedicated work. For a miniproject, which was required to be "double-marked" (with the grades from different people subsequently reconciled to determine a final grade) because of its open-ended nature, I asked the students to write a short paper on a specified advanced topic on networks. I changed the focal topic from year to year, and I show the miniproject that I assigned to the MFoCS students in my course in 2016 in Fig. 1. My goal was

---

[4] These special topics were marked both by at least one other person and by me (so-called "double marking"), and a reconciled mark from those scores became the student's grade in my course.

[5] The MFoCS program is a joint venture between the MI and the Department of Computer Science.

[6] I sometimes assigned papers that I knew well. Other times, I used the refereeing assignments as an excuse to carefully read a paper that interested me (and which, in practice, I otherwise might not read).

The University of Oxford

MSc (Mathematics and Foundations of Computer Science)

Networks (C5.4)

Hilary Term 2016

Below is listed a broad topic. Write a report on a specific subtopic within that general heading. Your report must include at least some numerical simulations (which you produce) and must include salient discussions of modeling issues, random-graph ensembles, and empirical data.

- Spatial Networks

Your report should be in the format and style of an article for the journal *Proceedings of the National Academy of Sciences*, and the main text must be no more than 6 typeset pages and must use their LaTeX style files (a template and style files will be provided). The report must include all sections (abstract, significance statement, etc.) in papers published in that journal (2016 format of papers). It is permissible to include a section of Supplemental Information that shows additional figures and calculations. In your report, indicate explicitly which ideas are new and which come from existing sources, and use appropriate and explicit attributions for all references (which must include papers reporting original research) or anything else (e.g., including code and figures) from other sources.

[*You need not submit scripts for any code you produce, but you may include them as part of Supplemental Information if you wish.*]

[*Your report need not contain original research results, though you must use some original research papers (not just review articles or books) as resources.*]

Fig. 1 The miniproject that I assigned to the MFoCS students in my networks course in "Hilary term" (winter term) in 2016. Their final grade was based on this miniproject, which was marked by at least two people (one of which was me), and then a final grade arose from a process of reconciling these grades.

for the students to have a miniature research experience (though the MFoCS program also includes a several-month dissertation as its capstone) and to cover an advanced topic in depth. Each year, I chose a focal topic that went beyond the course lecture material. Sometimes this entailed going into more detail on a capstone topic from units (6) or (7); other times, I selected a topic (e.g., "spatial networks" or "multilayer networks") from unit (8), even though I did not cover it in lectures beyond making a few cursory comments. As I discuss in Sect. 3.4, some computation (and potentially a lot of computation) is very important for the miniprojects, and ensuring that students are prepared to do them can be challenging.

3.3 Stage 3: Fourth-Year Undergraduates and Masters Students

In 2014, fourth-year ("Part C") undergraduates were able to take my networks course for the first time. Because of Oxford's end-of-year examinations, this necessitated moving my networks course from the spring term to the winter term ("Hilary term"). Unlike in the USA, most undergraduate courses in the MI are designated for students from one specific year, and these students also have to be from the mathematical sciences.[7] In the process of converting my networks course to an undergraduate course (which MSc students could also take), I also needed to formalize details such as recommended prerequisites, learning outcomes, assessment, and so on.

I indicated my learning outcomes in Sect. 2, and my course overview in the 2015 MI undergraduate course booklet read as follows:

> This course aims to provide an introduction to network science, which can be used to study complex systems of interacting agents. Networks are interesting both mathematically and computationally, and they are pervasive in physics, biology, sociology, information science, and myriad other fields. The study of networks is one of the "rising stars" of scientific endeavors, and networks have become among the most important subjects for applied mathematicians to study. Most of the topics to be considered are active modern research areas.

As I mentioned in Sect. 2, the goal of my course is to survey networks from the perspective of mathematical modeling and to teach students knowledge and skills to help them read the current research literature.

To make my course available for as wide a variety of students as possible, I did not suggest any prerequisites beyond what all undergraduates majoring in (i.e., "reading," to use UK parlance) Mathematics (and Mathematics & Statistics) are required to take anyway. For example, in the MI's official description of my 2015 networks course, I wrote the following text for recommended prerequisites:

> None [in particular, C6.2a (Statistical Mechanics) is not required], though some intuition from modules like C6.2a, the Part B graph theory course, and probability courses (at the level that everybody has to take anyway) can be useful. However, everything is self-contained, and none of these courses are required. Some computational experience is also helpful, and ideas from linear algebra will certainly be helpful.

The reason that I brought up the statistical-mechanics course, which I also developed, was that my networks course was labeled as C6.2b at the time, and the numbering could lead one to believe erroneously that material from the C6.2a course was required.

I was purposefully vague in my phrasing of "computational experience" in the recommended prerequisites, and the MI's computation requirement for first-year students was in the process of changing. As I discuss in Sect. 4, students' prior

---

[7] My UCLA version of the course (see Sect. 3.5), which I taught for the first time in spring 2017, included students from multiple majors and undergraduates in their fourth, third, and second years. A benefit of including second-year and third-year students is that some of them may desire to do an undergraduate research project on networks, and taking a networks course sufficiently early may also influence the subsequent courses that they elect to take.

experience with computation is one of the main challenges of teaching my course. The linear algebra that is required for students who are reading Computer Science (or Mathematics & Computer Science) is somewhat different than that for other undergraduates in the mathematical sciences, but in practice this issue never came up (or at least it never came to my attention) in my networks course. The students did occasionally ask questions about concepts from linear algebra and probability (e.g., generating functions show up a lot) that are important for my course, and such questions have been even more prominent in the UCLA version of my course (see Sect. 3.5).

With undergraduates now taking my course, I also needed to develop more formal homework assignments. To discuss these assignments, the lectures were supplemented with six hours of problem classes. (In practice, the total amount of time is somewhat shorter than six hours.) I arranged these as four 1.5-hour classes in 2014 and 2015 and as six one-hour classes in 2016. Problem classes are like the recitation sessions (sometimes called "discussion sessions") in US universities—although the UK problem classes are arguably structured more around homework assignments than is the case in the USA—and they normally are attended by undergraduates, MFoCS students, and students in the Mathematical and Theoretical Physics (MTP) program.[8] In problem classes, a "tutor" (who is in charge of one or more sets of classes), with some help from a teacher's assistant (TA), goes through homework problems that students find difficult, discusses reading assignments and any papers for which the students are supposed to write referee reports, walks through bits of code for computational exercises, and so on. I was a tutor for some sets of classes that were associated with my lectures, and postdocs or senior PhD students were tutors for other sets of classes.

Initially, as is standard in the MI, undergraduates received a grade for my course based entirely on one exam that they took at the end of the academic year. The fourth-year students in the mathematical sciences start having their exams in the middle of Trinity term (and hence starting around the end of May). Homework problems and other materials are meant to help undergraduate students learn and prepare for a final exam, but any "grades" on assignments are intended only for feedback; they do not affect the course grade. My homework assignments were a mix of problems that I hoped would help students prepare for the exam and longer (and occasionally open-ended) problems to encourage them to explore topics in detail in a way that is impossible in an exam question.

My course's exam lasted for 1.5–1.75 hours (it varied because of rule changes) and included three problems. The students received a grade based on their top-two marks among those problems. Because of this setup, which I inherited from MI rules, many students choose one problem to skip (sometimes based on course material that they had decided that they would not bother studying in detail) and focus their efforts on the other two problems. Because of this mechanism, people who write exams often try to make problems of equal difficulty, a very time-consuming

---

[8] Starting in Hilary 2016, students from Oxford's initial cohort of a new MSc program (which I helped design) in Mathematical and Theoretical Physics could also take my course.

and essentially impossible task, given that sometimes different topics have inherently different difficulty levels.

Assessment of the masters students followed the norms for their various programs. The MFoCS students were required to do the standard homework assignments in addition to their refereeing homework assignments, and they were assessed by miniproject at the end of Hilary term (see Sect. 3.2). The MMSC students could choose to do a special topic in my networks course (as one of the set of special topics that they are required to do for the program) if they wanted to receive a grade in it (see Sect. 3.1). The MTP students were required to do the same miniproject as the MFoCS students.

3.4 Stage 4: Changing from Exam Assessment to "Miniproject" Assessment

The final major change in my networks course at University of Oxford was converting the undergraduate assessment from exams to miniprojects. (See my discussion at the end of Sect. 3.3.) I taught the 2016 version of my course with miniproject-based assessment.

In my view, examination-based assessment is particularly inappropriate for a course about networks. Problems in this format are artificially short and depart substantially in both scope and time allotted from the types of problems that one actually studies in network science. Thus, although I used exam-based assessment voluntarily during the first year that undergraduates could take my course (see Sect. 3.3), I did so with the expectation of changing it shortly thereafter. After a long and (very) tedious battle, I was able to convince the MI's teaching committee to allow this change in 2015, which allowed me to implement it in 2016.[9]

Assessing the undergraduates in my course using miniprojects, which I was already doing for MFoCS students (see Fig. 1 for an example), gave them an opportunity to explore a topic in depth and provided an introduction to doing research in network science. The benefits of using miniprojects for teaching students about networks also hold at other levels, as demonstrated by the NetSci High program for teaching network science to high-school students [24].

Although the miniprojects that I used for the undergraduates in my course closely resembled the ones that I was already using for the MFoCS students, there were a couple of important differences. First, instead of picking one broad topic for the students, as I did with the MSc students, I gave undergraduate students a choice between two broad topics—"community structure and other mesoscale structures in networks," which goes predominantly with unit (6), and "spreading processes on networks," which goes predominantly with unit (7)—partly because I wanted them

---

[9] I believe that my course was the first lecture-based course in the MI to be approved for miniproject-based assessment. It was the first domino to fall, and at least one other course soon followed suit. I expect that there will be more.

to have some choice and partly because I wanted to give myself a bit more variety, given that I was going to be evaluating more than two-dozen reports. The other instructions in the miniproject (again see Fig. 1) were the same for both undergraduate and MSc students. Second, I purposely connected the project topics directly with capstone subjects in the course, whereas I was a bit more adventurous with the MSc students, who I felt should spend time on a topic that itself went beyond what was in the course.

Using miniproject-based assessment necessitated some tricky changes in timing. To the extent possible, the MFoCS miniprojects (which I also used for the MTP students) and undergraduate miniprojects needed to be synchronized—and both types of projects were to be undertaken during a 3-week window, with an expected commitment of 3–4 days of strenuous work—so the undergraduate miniprojects needed to occur at the end of Hilary term (as that time was fixed for the MSc students), rather than in the middle of Trinity term. For the undergraduates, we released the miniproject on Monday of the eighth and final week of Hilary term. The sixth and last problem class could thus occur no later than during the seventh week; this enforced more rigid timing at the end of my course than was the case when assessment was based on an exam to be taken a few months later. Grading so many projects (about three dozen, counting all students) was rather strenuous and time-consuming, and some of the grade reconciliation with the other markers was highly nontrivial. On the bright side, I didn't need to grade any exams or spend dozens of hours constructing an exam.

Importantly, the change from exam-based assessment to miniproject-based assessment gave me much more freedom to teach my course in the way that I wanted. I made my homework assignments "more realistic" with respect to what practitioners of network science do in their research. Even before the change, my homework assignments included several problems that allow exploration, a significant departure from the norm in the MI. After the change, I further reduced focus on problems of a style that align with exam preparation, and I increased emphasis on computation, as this is a very important aspect of network science. Practicing computational explorations also helps prepare students for undertaking a miniproject.[10] Another of my changes was to reduce the number of problems with similar calculations, such as generating-function analyses with progressively more intricate random-graph models, as I wanted students to see a small number of examples to get an idea about methods, rather than overemphasizing some topics at the expense of others. I also added some "refereeing" problems (though many of the undergrads seemed to struggle with these, or at least were perplexed by them), like the ones that I had already been assigning for several years to the MSc students (see Sect. 3.2). The MFoCS students needed to do both these refereeing problems and the ones that were designed specifically for them.

---

[10] Even before changing my course's mode of assessment, my homework assignments included some computational exercises, which many students tended to ignore, perhaps because they thought (mistakenly) that I couldn't test the material in these problems on timed exams without computers.

My revamped course worked much better than the examination-assessed version—which already worked much better the second time that it was offered it to undergrads, as there were many kinks to iron out from the first year—and I think that most of my students agreed with me. The MI ended up giving me a teaching award in recognition of designing and teaching my networks course.

The networks course still exists at University of Oxford, and it was taught by Heather Harrington in 2017 and by Renaud Lambiotte in 2018. Even though the course is only a few years old, it was already well-attended the first time that it was offered to undergraduate students, and in 2017 more undergraduates were enrolled in it than in any other fourth-year mathematics course at Oxford.

3.5 Stage 5: Transferring My Course to UCLA

In 2016, I moved to the Department of Mathematics at UCLA, and I taught a version of my undergraduate networks course in spring 2017. I taught it as a special-topics course, and it joined the course catalog as a regular offering starting in the 2017–2018 academic year.

For the initial UCLA version of my course, I mostly followed what I had been doing at Oxford, as I wanted to see what transpired in practice to have a better understanding of what, if any, substantive things needed changes. (I'd rather make such changes once rather than twice, and I felt that the existing form of the course was rather good.) Given that I moved back to the USA, I also went back to having formal office hours, which doesn't occur for courses at Oxford. Naturally, I am also available by appointment and answer queries by e-mail and on the course discussion board.

The extra freedom in the US system compared to Oxford allowed me to make a few formatting changes, such as in how I assess students and determine their grades. My class still has miniprojects, though I decided to make them group projects, allowing students to go further with them and making it more manageable to grade them. The miniprojects constitute 50% of the course grade, and there is both a group written report (in the format of the journal *Proceedings of the National Academy of Sciences*, as I was doing with the course at Oxford) and a group oral presentation. The homework assignments make up 25% of the grade, and quizzes and one midterm (where the midterm, which the students take during one 50-minute class period, counts as three quizzes) accounts for the final 25% of the grade.

In the spring-2017 offering of my course, there were 3–4 students in each miniproject group, and I determined the groups a few weeks into the course after soliciting ideas from the students about what topic they might want to study (though without a commitment to a specific topic) and with whom they might wish to work. In practice, with various other course commitments (such as homework assignments), the students had about three weeks to do their miniprojects. As with a PhD-student-level networks course that I taught during the winter-2017 term, I met with each group of students to help get them started in the early stages of their miniprojects (e.g., to make sure their project was doable, such as by ensuring that they were not using data that would take much longer than the three-week project timescale to clean before they could do analysis), and I was also available for consultation about their miniprojects throughout the time that they were working on them.

With 10 weeks rather than 8 weeks in a term and with three scheduled lectures each week (except for a couple of holidays), a UCLA course has a lot more lecture time than the ones that I taught at Oxford, and there is also a one-hour (technically 50 minutes) recitation section once a week. There is a single weekly discussion session for all students (in spring 2017, there were about 23 of them) in my course, in contrast to Oxford, where my course had multiple such sessions (with about 8–12 students each). I haven't added new mathematical material to the course, and I have instead used the extra time to add introductory discussions to cover the "big picture" in both complex systems and network science, go through some of the material more slowly, and interact more closely with the students in lectures. Having only 16 lecture hours at Oxford rushed things, and even student questions in lectures typically couldn't get the attention that they deserved. Moreover, when I was using exam-based assessment—and I was required to submit exam materials extremely early, entailing a strict commitment to cover the material that was being tested—I had almost no leeway to veer away from the course's intended trajectory, and not getting through the necessary material was simply not an option. At UCLA, in contrast, if one doesn't cover a topic, one can just not put it on an exam. Moreover, my networks course is an elective, and I don't need to worry about covering material that is prerequisite for other courses.

As was the case at University of Oxford, my homework assignments at UCLA include a mixture of straightforward problems that are meant to help the students learn definitions and concepts, trickier problems to stretch their knowledge about them, and computational exercises (including open-ended ones). Because I no longer have exclusive responsibility for working out detailed solutions of the homework problems (TAs help with this at UCLA), I assign several problems from [1], unlike what I did at Oxford. When I first taught my networks course at UCLA, I intended to again include paper-refereeing problems, though I didn't do so in practice; I hope to include them sometimes in future years. I also created a couple of homework "problems" of an unusual nature. For example, as part of the first homework assignment, I ask the students to take a picture of a local network (either on campus or in Westwood, which is an area next to campus), to identify the nodes and edges, and to indicate any other features that they notice (e.g., whether it is a spatial network, has multiple types of edges, and so on). My hope with this problem is to encourage my students to think about the fact that networks are everywhere.

All of my quizzes at UCLA have been "pop" quizzes (lasting about 15–20 minutes), and there were three of them in total in my spring-2017 course, though I did not fix the number before the term started. The main purpose of the quizzes— and especially of not announcing in advance when they are going to occur—is to encourage the students to attend lectures and to keep up with the homework and the reading. Similarly, the main purpose of the midterm (for which the students can use hard copies of their lecture notes, their homework assignments, and [1]) is to encourage students to spend time poring over the material to learn it better.

Unsurprisingly, the student composition of my networks course has been rather different at UCLA than it was at Oxford. At UCLA, most of the enrolled students are majoring in the mathematical sciences, though there are some exceptions, and there are now third-year and second-year students in addition to fourth-year students. Additionally, my course now includes only undergraduates. I now need to list recommended prerequisites—which are appropriate linear-algebra and probability courses, along with the desirability of some prior experience with programming—as it is no longer guaranteed that students who want to enroll in my course have previously seen certain essential topics. As was the case at University of Oxford, some relevant topics (such as generating functions) aren't covered in the prerequisite courses at UCLA, so I introduce them myself and encourage the students to look them up in detail on their own if they want more information. The level of prior programming experience (and degree of difficulty in getting started with the computational exercises) is mixed among the students, much like what I had observed at Oxford, and the command of linear algebra among my UCLA students is weaker overall than was the case for my Oxford students. I have allowed students to take my networks course even if they haven't taken courses in the prerequisite subjects. The point of specifying those subjects is to convey what knowledge I am going to assume from the first day of my course.

I expected some hiccups in my course from the institution change and the ensuing differences in its composition of students, but its UCLA debut in spring 2017 was unexpectedly smooth. My course was very popular among the students who took it—I received even more positive course evaluations than the ones from the 2016 course at Oxford—and it benefited a great deal from the extra lecture time (especially from not having to rush things and being able to interact a lot more with the students), having office hours, and other things. There were just over 20 students in my course, and learning about networks appears to have inspired them: several of them are collaborating with me on research projects, and two others contacted me to let me know that they were using skills and knowledge from my networks course in their job internships. One comment from the students that I have implemented for my course in 2018 is to start the projects earlier, which had been my intention last year. The delayed time before starting project work arose from wanting to show the students enough topics so that they would have a better idea of what they might want to work on and how to go about doing it, and presenting this amount of material took longer than I expected. I like the pace at which I can now present the course material, and I have tried to preserve that while also introducing project work earlier to give the students a bit more time on their projects.

4 Some Challenges

Teaching my networks course has been very challenging. Key challenges have included (1) finding the right balance, especially given the diverse backgrounds of my students, between mathematical rigor and models, methods, and using a physical-applied-mathematics approach; (2) computational exercises and expectations, in conjunction with diverse student backgrounds in computation and programming; and (3) exam-based assessment (until I was able to change this).

The diversity of the mathematical backgrounds of the students in my course has been a persistent challenge. For example, in the University of Oxford version, most students had one of two principal backgrounds: (1) people who had taken many applied-mathematics courses (e.g., in topics like fluid mechanics, differential equations, and mathematical biology) but who had taken few or no advanced courses in pure mathematics, where the focus is on mathematically rigorous arguments, and who were also not used to applying ideas from "physical applied mathematics" to discrete structures; and (2) people who had taken a lot of courses in discrete mathematics (in topics such as graph theory, probability, and various areas of statistics) who were more comfortable with mathematically rigorous arguments and/or statistical modeling than with doing physical-applied-mathematics modeling [25, 26] using arguments that usually are not mathematically rigorous.

My approach to studying networks—in both teaching and research—follows the tradition of physical applied mathematics [2], which emphasizes modeling and scientific rigor (and domain relevance) but typically does not focus on demanding mathematical rigor. In my networks course, I put strong emphasis on mechanistic modeling, but I usually sacrifice mathematical rigor (especially given the time constraints), and I almost never present things in a precise definition–example–theorem format. Moreover, many topics that I discuss would take a very long time to present in a mathematically rigorous way or are not yet even known at that level. I discuss much more complicated models than what one typically sees in a graph-theory course (or in a rigorous statistics course on networks),[11] and I discuss the application of (mechanistic) modeling principles that are taught much more commonly in applied-mathematics courses than in statistics (where descriptive modeling is emphasized) or pure-mathematics courses.

It is very challenging to cover formal definitions and theory, and then to discuss dynamics and modeling, and then to relate them to real data sets and numerical computations. This challenge was already present when only MSc students were taking my networks course, as MMSC students are mostly of type (1), but MFOCS students are predominantly of type (2); and it became even more prominent when I was also teaching undergraduate students. In my course, I occasionally bring up some physics jargon (much of which is used in [1]) to help make connections (e.g., to some topics in statistical mechanics), though I try to do so without overemphasizing it. Because some of the classical models in network science are not

---

[11] For example, in graph theory, one might spend a lot of time rigorously proving results on Erdős–Rényi (ER) random graphs, but I want to spend time on more intricate random-graph models (such as configuration models and their generalizations) that are more appropriate for studying real-world networks. I do introduce ER graphs in my course, and various homework problems are about them, but I discuss heuristic arguments for analyzing them, rather than presenting mathematically rigorous arguments (which carries the risk of drowning students in details), to demonstrate important ideas (such as a phase transition to a giant connected component).

well-defined mathematically in most of the standard presentations of them,[12] it is easy to fall into a trap in the middle of a lecture of not specifying everything that is necessary in a manner that is sufficiently precise. This was especially frustrating to Oxford students with a pure-mathematics background, as most of them are not used to this style of presentation. I try to be more precise in my presentation of network models than is often the case in research papers, but it is rather challenging to strike the right balance between precision of model specifications—as well as the level of mathematical rigor when analyzing the models, especially given that there are many features of these models for which mathematically rigorous analysis remains an open challenge—and an emphasis on modeling and discussing examples of many different types of models. Thankfully, when students examine these network models computationally on homework assignments and in their miniprojects, they get a chance to see (and, ideally, discover for themselves) exactly what information is needed to ensure a complete, precise specification. Moreover, investigating the consequences of making different choices in a given family of models is an important topic to study. (I like to include homework problems that encourage such exploration.)

A second major challenge, which occurred both at University of Oxford and at UCLA, is the diversity in students' past experiences and knowledge about doing computations. When I assign computational exercises as part of homework—note that it is very hard to make these problems equally accessible to students of widely differing computational backgrounds—I go out of my way to ensure that code (e.g., in MATLAB, for which University of Oxford has a site license) is available online, such as through the Brain Connectivity Toolbox [27] or other resources. Several of my homework problems and the course miniprojects require the use of real-world data sets, and many students stumble upon some of the famous (and infamous) data sets, such as ones that are available from Mark Newman's website [23]. Eventually, I started including some tutorial computational exercises at the beginning of my course. This helps a lot, but it has not completely removed the challenges for students with less computational experience or knowledge. Starting with release R2016a, MATLAB has included some functionality for network analysis [28], and this has been helpful for my course. Some of my students have decided that they prefer using Gephi [29] (e.g., for visualization), Python with NetworkX [30], or R with igraph [31]. I am happy with the students using whatever software they want.

My concern for my course is not whether the undergraduates can program in MATLAB or using any other language or software package, but rather that they can successfully use, understand, and interpret the output of computations. If that means running somebody else's .m file in MATLAB, so be it. Some programming experience does help, but strictly speaking it has never been something that my course has required (despite the feelings of some students to the contrary). One frustrating issue, especially for students, that sometimes arose at Oxford is that students may not know where to go if they cannot get code or a software package to work. In US

---

[12]For example, one needs to make choices for how one rewires or adds shortcuts in a Watts–Strogatz network [32, 33], one needs an initial ("seed") network when studying preferential-attachment models [1], and so on.

universities, such issues tend to be less problematic, as the office hours of professors and teaching assistants are great for addressing these kinds of individual queries. There were fewer opportunities at Oxford to help students sort through such technical problems (which often are not easy to address with e-mail communications), and my TAs and I encouraged students to talk to other students who had managed to get a particular package to work or to look things up online. As I mentioned previously, when my networks course was assessed by a final exam, many students ignored the computational exercises on the homework assignments, as they seemed to think that they weren't testable (despite my explicit comments to the contrary). However, that is mistaken, as an exam can include questions that describe or show output, and I can then ask the students about it.

Another issue that is worth bringing up is my course's reading assignments. In mathematics courses at Oxford, lecturers are not allowed to compel students to buy any textbooks for courses—a marked difference from the norm in US universities—so it is standard to provide students with a terse set of lecture notes. I wanted my students to read material beyond what was in my notes. In early versions of my course, in addition to a scanned version of my notes (I received complaints that they weren't typeset), I gave the students a copy of an in-progress textbook in its very rough state, and I made it clear that it was very far from polished. I also strongly encouraged my students to go through various parts of [1], as well as other resources (such as parts of some review articles), though they did not always find it clear which source they should use for a given topic. Mathematics undergrads at Oxford tend to focus on material in lecture notes, and my MI courses were unusual in expecting much more reading than what is in a short set of notes. Naturally, there is far more to the material in an advanced course than what is included in a terse set of notes. Such reading was optional, though I strongly encouraged it, through the 2014 version of my course, because I was unable to ensure that all students had easy access to [1] without forcing them or their Colleges (each Oxford student is a member of a College) to buy the book.

A few months into 2014, the first year that my course was open to undergraduates, I found out from one of my students that all Oxford students could freely access [1] online, so starting in 2015 I assigned explicit reading from [1] and elsewhere as the first "problem" on every homework sheet. I thereby informed students exactly what I required them to read, and I also suggested some optional additional reading that I felt would be helpful. This largely solved the problem of what the students should read—and it helped the 2015 version of my course to work out *a lot* better than the 2014 version, which was extremely rough—though some students complained that there was too much material to read. Others felt that [1] is fast and easy to read and that reading about 50 pages per week of it is very manageable. For the UCLA version of my course, I require my students to buy [1] (but only that book), and I continue to use sources like review articles and other online resources (as well as my lecture notes). From my experience teaching undergraduates at University of Oxford, I think that too many of them seem to prefer the boring Oxford model of lectures and exams, with "examinable" material specified in a terse set of lecture notes, and I purposely (and purposefully) taught my networks course in a more exotic way. I think that my adventuresome approach greatly benefits the students in my courses, even if some of them are not always happy about it.

5   Conclusions

When I was at University of Oxford, I developed an introductory course in network analysis that is now taken by numerous fourth-year undergraduates and masters students from several programs in the mathematical sciences. I have also translated this course into one for undergraduates (at "upper division" level) that I teach in the mathematics department at UCLA. In both variants, my networks course links ideas from applied mathematics, theoretical (i.e., "pure") mathematics, and computation through the modeling and investigation of discrete structures. An introductory mathematics course about networks is an important component of a "discrete structures and data science" pathway through an undergraduate degree in the mathematical sciences, and it is crucial for universities to include these types of pathways.

I hope that the present article will help encourage faculty—especially those in mathematics and mathematical-science departments—at other institutions to design and teach introductory courses in network analysis. My course is for advanced undergraduates, and it would also be good to develop courses in network analysis (e.g., freshman seminars) that are appropriate at an earlier stage of undergraduate education. Such courses complement existing courses in graph theory and other subjects, and they give a chance to introduce students to state-of-the-art topics that apply ideas from graph theory, probability, dynamical systems, and other important subjects in fascinating ways. I have suggested topics that are appropriate for an introductory networks course, and I have strongly advocated the use of miniprojects as a key method of assessment. As I have discussed, there are various challenges (e.g., diverse computational and coursework backgrounds among the students) in teaching an introductory course on networks, but it is a very valuable offering, and every mathematical-science department should include one. I hope that my description of my experiences will encourage the development of more undergraduate-level courses on networks in mathematics programs.

Acknowledgements   I am grateful to the various "tutors" and teaching assistants who have helped me immensely when I have taught my networks course at University of Oxford and UCLA. I thank Heather Harrington, Florian Klimm, Flora Meng, Roxana Pamfil, Steve Strogatz, and Fabian Ying for helpful comments. I particularly thank Hiroki Sayama for reviewing an earlier version of this manuscript and providing very helpful feedback. (I also thank Hiroki for his support when dealing with the typesetting issues in this chapter.)

References

1. Newman, M. E. J. (2010) Networks: An Introduction, Oxford University Press.
2. Porter, M. A. and Howison, S. D. (2017) The role of network analysis in industrial and applied mathematics. ArXiv:1703.06843.
3. Society for Industrial and Applied Mathematics (2017) Careers in Applied Mathematics: Options for STEM Majors. https://www.siam.org/careers/thinking/pdf/brochure.pdf, 2017. Accessed, 1/19/18.

4. Porter, M. A. (2016) Math C5.4: Networks. http://networksoxford.blogspot.co.uk. Accessed 1/19/18.
5. Briatte, F. (2017) Awesome network analysis: Courses. https://github.com/briatte/awesome-network-analysis#courses. Accessed 1/19/18.
6. Clauset, A., (2017) Network Analysis and Modeling, CSCI 5352. http://tuvalu.santafe.edu/_aaronc/courses/5352/. Accessed 1/19/18.
7. Donner, R. and Kurths, J. (2014) Lecture "Complex Networks" — Winter term 2014/15. https://www.pik-potsdam.de/members/redonner/lecture-complex-networks-summer-term-2014. Accessed 1/19/18.
8. Dovrolis, C. (2017) CS 7280: Network Science: Methods and Applications. http://www.cc.gatech.edu/_dovrolis/Courses/NetSci/. Accessed 1/19/18.
9. D'Souza, R. (2016) ECS 253 / MAE 253: Network Theory and Applications. http://mae.engr.ucdavis.edu/dsouza/ecs253. Accessed 1/19/18.
10. Easley, D. and Kleinberg, J. , Economics 2040 / Sociology 2090 / Computer Science 2850 / Information Science 2040. https://courses.cit.cornell.edu/info2040_2014fa/. Accessed 1/19/18.
11. Giannini, M. Network Science PhD Program @ Northeastern. https://www.networksciencein-stitute.org/phd (click on 'Courses'). Accessed 1/19/18.
12. Jaillet, P. (2017) 6.268 Network Science and Models (Spring 2017). https://stellar.mit.edu/S/course/6/sp17/6.268/index.html. Accessed 1/19/18.
13. Kearns, M. (2017) Networked and Social Systems Engineering (NETS) 112: Networked Life. http://www.cis.upenn.edu/_mkearns/teaching/NetworkedLife/. Accessed 1/19/18.
14. Newman, M. E. J. (2015) Complex Systems 535/Physics 508: Network Theory. http://www-personal.umich.edu/~mejn/courses/2015/cscs535/index.html. Accessed 1/19/18.
15. Salganik, M. (2017) Sociology 543: Social Network Analysis (Spring 2017). http://www.princeton.edu/_mjs3/soc543_s2017/. Accessed 1/19/18.
16. Santa Fe Institute (2018) Complexity explorer: Courses. https://www.complexityexplorer.org/courses. Accessed 1/19/18.
17. Shalizi, C. R. (2016) 36–720, Statistical Network Models: Mini-semester I, Fall 2016. http://www.stat.cmu.edu/~cshalizi/networks/16-1/. Accessed 1/19/18.
18. Wierman, A. (2018) CS/EE/CMS 144: Networks: Structure & Economics. http://courses.cms.caltech.edu/cs144/. Accessed 1/19/18.
19. Hiroki Sayama. (2018). Mapping the curricular structure and contents of network science courses. In: Catherine B. Cramer, Mason A. Porter, Hiroki Sayama, Lori Sheetz, Stephen Miles Uzzo, eds., Network Science In Education: Transformational Approaches in Teaching and Learning. Cham, Switzerland: Springer International Publishing (arXiv:1707.09570). This article appears as Chapter 7 in this book.
20. Fortunato, S. (2010) Community detection in graphs, Physics Reports 486: 75.
21. Porter, M. A., Onnela, J.-P., and Mucha, P. J. (2009) Communities in Networks. Notices of the American Mathematical Society, 56: 1082.
22. Good, B. H., de Montjoye, Y. A., and Clauset, A. (2010) Performance of modularity maximization in practical contexts, Physical Review E, 81: 046106.
23. Newman, M. E. J. Network Data. http://www-personal.umich.edu/~mejn/netdata/. Accessed 1/19/18.
24. Cramer, C., Sheetz, L., Sayama, H., Trunfio, P., Stanley, H. E., and Uzzo, S. M. (2015) NetSci High: Bringing network science research to high schools (pp. 209–218), in Complex Networks VI: Proceedings of the 6th Workshop on Complex Networks (CompleNet 2015), G. Mangioni, F. Simini, S. M. Uzzo, and D. Wang, eds., Studies in Computational Intelligence 597, Cham, Switzerland: Springer International Publishing (arXiv:1412.3125).
25. Fowler, A. C. (1997) *Mathematical Models in the Applied Sciences*, Cambridge, UK: Cambridge University Press.
26. Lin, C. C. and Segal. L. A. (1988) *Mathematics Applied to Deterministic Problems in the Natural Sciences, Classics in Applied Mathematics*. Philadelphia, USA: Society for Industrial and Applied Mathematics.

27. Rubinov, M. and Sporns, O. (2010) Complex network measures of brain connectivity: Uses and interpretations, NeuroImage, 52: 1059. Code is available at https://sites.google.com/site/bctnet/. Accessed 1/19/18.
28. MATLAB: Graph and Network Algorithms. https://uk.mathworks.com/help/matlab/graph-and-network-algorithms.html. Accessed 1/19/18.
29. Gephi — The Open Graph Viz Platform. Available at https://gephi.org. Accessed 1/19/18.
30. NetworkX — NetworkX. Available at https://networkx.github.io. Accessed 1/19/18.
31. Igraph – Network Analysis Software. Available at http://igraph.org. Accessed 1/19/18.
32. Porter, M. A. (2012) Small-world networks, Scholarpedia, 7. 1739.
33. Watts, D. J. and Strogatz, S. H. (1998) Collective dynamics of 'small-world' networks, Nature, 393: 440.

Leading Edge Learning in Network Science

Ralucca Gera

1 Introduction

The current work describes the method that has been used since 2014 at a the Naval Postgraduate School (NPS) in teaching MA 4404, the Structure and Analysis of Complex Networks course, primarily to USA and international officers during their master's and doctoral program. This course is taught in the Applied Mathematics Department, as part of the Network Science Academic Certificate that students can receive along with their master's or doctoral degrees in any math-related curriculum. The students interested in the course have a technical background, generally in mathematics, computer science, operations research, or engineering. Additional information on this course and how it fits within the certificate can be found at http://faculty.nps.edu/rgera/NetSci/Certificate/dist/index.html.

Researchers have taught and used network analysis since the eighteenth century, starting with the classic Seven Bridges of Königsberg problem in graph theory [31]. In many mathematics courses, students in the same curricula with the same background and interest were exposed to information building on the same mathematical prerequisites. In recent years, students with mixed backgrounds want to learn about networks, which initiated the desire to modify standard teaching, from motivation to solutions and their interpretations, in particular, taking on a guided discovery approach, asking students to experiment with networks, and discover the reasons behind what they observed, in order to support student learning (which sometimes is hard to optimally use) [18, 27, 41].

As I promote active learning in teaching, defined by Bonewell et al. [17], the challenge is to undertake learning activities that will engage students and motivate their learning, regardless of the individual background. This stimulates interest in the class, improves the learning experience, and increases their chance of successfully

R. Gera (✉)
Naval Postgraduate School, Monterey, CA, USA

© Springer International Publishing AG, part of Springer Nature 2018
C. B. Cramer et al. (eds.), *Network Science In Education*,
https://doi.org/10.1007/978-3-319-77237-0_2

recalling and using learned ideas in the future [13, 33]. As student interest in a course increases if the course is relevant to the student, I make it pertinent by allowing students to pick individual networks that they will analyze for the rest of the course. As I teach new concepts, they apply them to this chosen network, making it both interesting to them and to the rest of the class as they see a variety of networks analyzed. As the analysis of these networks is discussed in the class, it brings the contrast and diversity needed for learning.

Since I had the freedom of designing the first course in network science at NPS, I made the following choices, which may change. The analysis tools chosen for the course incorporate choices for each type of student background: Gephi [1], Python [2], and R [3], commonly used in network science [10, 20, 24, 39, 45]. Most students use Gephi for visualization and either Python or R for analysis and constructions. A Research Project (Sect. 5.3) is incorporated in the course, as I believe that project-based learning is a good approach to education designed to engage students [15]. Moreover, through the design of the Research Project (Sect. 5.3), students learn $L^A T_E X$, most of them typing their theses in $L^A T_E X$ in the quarters following the current one. The last main component of the class (and of the grade as well) is the alternative to standard mathematical homework assignments: the Network Profile Summary (Sect. 5.4) in which students apply weekly learned concepts to analyze a personal network, rather than all students analyzing the same network/ data. I believe it promotes self-determination and confidence in the learned topics, and it has been shown that self-determination motivates student learning [28, 52].

The course generally starts by showing existing networks, as well as how big data can be modeled by networks. Currently this is of interest to most of my officers in trying to understand emerging phenomena in technology and society. Some examples of networks presented in class are online social networks, the Internet, the World Wide Web, neural networks, food webs, metabolic networks, power grids, airline networks, national highway networks, the brain, and others. These examples are complemented by networks that students choose to create based on their interests and previous experience, including: terrorist networks, the US Tesla network, the global transportation network, snapshots from YouTube, and Twitter data.

The course then proceeds with the basic generative models (random graphs, small-worlds, preferential attachment) and newer ones based on the interest of the class. I start with the Erdős–Rényi random networks that combine the just-learned concepts of graph theory with probability theory, followed by more sophisticated models of network formation, including: Milgram's 1967 experiment [53] and Watts–Strogatz small-world networks [55], the Barabási–Albert preferential attachment growing model [9] and its variants, the Molloy-Reed configuration model [47], the random geometric model [29], and other ones that are relevant at that time. In the last couple of years, the Research Project (Sect. 5.3) has built on this overview of generative models by asking students to create new ones.

Once armed with the real and synthetic networks examples, the rest of the course focuses on analysis of networks. The topics covered are not consistent from year to year. As this field is evolving so quickly, one of the goals for this course has been not to use a static book for teaching but rather to present slides and research articles based on exposure to ideas presented at the most recent conferences, such as NetSci (https://www.netsci2018.com/), CompleNet (http://complenet.weebly.com/), SIAM Workshop on Network Science (http://www.siam.org/meetings/ns18/), ASONAM (http://asonam.cpsc.ucalgary.ca/2018/), and Sunbelt (https://sunbelt.sites.uu. nl/). Before coming to my lectures, students have to watch TED talks on the topics that I will be teaching that day. This way, they have a different point of view of why the topic is interesting, and how other researchers have used it. They then come to my class with questions for me, which allows them to hear the answers I have on what I teach that day. The presentation slides are updated regularly, exposing the students to updated information and from several sources. It is important to me that students get multiple points of view on the topics, since network science researchers come from different fields of studies, emphasising the motivation and application they found. The slides are complimented with articles and recorded lectures from conferences and one or two standard books for references.

The rest of this chapter is organized as follows. Section 2 presents the course learning outcomes and objectives, followed by the content and software used in the class and detailed in Section 3. Section 4 gives the overview of the course format, detailed in Section 5, as well as including the assessment for each activity. Finally there is a conclusion and student feedback.

2 Course Learning Outcomes and Objectives

The goal for students in this class is to develop the mathematical sophistication needed to apply learned methodologies to, and understand properties of new networks. This course enables them to have enough exposure and practice to readily use existing concepts or further read and understand published research as needed for future projects.

To achieve this, students analyze their personal network for the Network Profile Summary by practicing the introduced concepts of complex network analysis and by describing the structure of the chosen network. Furthermore, they contrast network models to real networks, by explaining features some complex networks have that others do not. This allows them to synthesize the new research in this evolving area and critique a peer's research. Students also read papers for which they have to grasp and explain new research ideas in complex networks.

The outcome of the course is that through new network research, the students will design new network models building on existing ones and available data. They will be able to design experiments to test hypotheses based on analyzed data and generate new methodologies by expanding on the designed experiments.

The learning outcomes above are achieved through building and analyzing personalized networks, reading scientific papers, writing technical research articles for publication, and presenting them at network science conferences. In my view, these are exactly the puzzle pieces for attaining the learning objective of the course: understanding the concepts, models, and methodologies needed to identify how to use knowledge of complex networks to produce a research article or apply in a real-world situation. This gives students the mathematical sophistication and confidence to use gained experience as situations arise.

3 Course Content and Software

The course materials are available at http://faculty.nps.edu/rgera/MA4404.html. The topics of the course are the following:

1. Types of networks:
2. Synthetic network models: Erdős–Rényi random networks, Watts-Strogatz small-world networks, the Barabási–Albert preferential attachment growing model and its variants, the Malloy-Reed configuration model, and the random geometric model;
3. Network statistics/properties: degree, clustering coefficient, diameter, density, shortest paths, node similarity, and homophily; and
4. Centralities: degree, closeness, betweenness, eigenvector, Katz, PageRank, hubs, and authorities.

These topics get augmented by presentations based on information from current conferences. Because of the fragmented literature–with inconsistent terminology and frequent reinvention of concepts and methodologies of network science due to the mix of the backgrounds of their researchers–this class builds on several manuscripts and conference presentations. Presentation slides are available for each lecture day at http://faculty.nps.edu/rgera/MA4404.html, and they are updated based on new research and information and animations from conferences. The main articles used as the class references are (a) Newman's 2003 article "The Structure and Function of complex networks" [48] which can be found at http://epubs.siam.org/doi/pdf/10.1137/S003614450342480 from SIAM and (b) Barabási–Albert's free and interactive book *Network Science* which can be found on his website at http://barabasi.com/networksciencebook (and also in print [8]).

The main visualization tool is the open-source graph visualization and manipulation software Gephi, found at http://gephi.org/ [1]. To complement Gephi's analysis ability, I generally use Python or R, open-source programming languages with wide interoperability, and other tools [10].

For Python users https://www.python.org/ [2], I suggest NetworkX (https://networkx.github.io/) and igraph (http://igraph.org/redirect.html), two Python libraries developed for the study of graphs and networks.

For R users https://www.r-project.org/ [3], I suggest igraph, http://kateto.net/networks-r-igraph and an overview found here (http://www.necsi.edu/events/iccs6/papers/c1602a3c126ba822d0bc4293371c.pdf) [24], or Statnet with the following tutorial from a Sunbelt conference: https://statnet.org/trac/raw-attachment/wiki/Resources/introToSNAinRsunbelt2012tutorial.pdf.

Available tools are updated on a regular basis, as this information is not static: http://faculty.nps.edu/rgera/MA4404.html

4 Course Format

Classes meet Monday through Thursday, for 50 min each. For the first 3 weeks of the quarter, interactive lectures are provided for each class. During this time, students are exposed to an overview of network science and real and synthetic networks.

There are two assignments due in these 3 weeks. The first assignment is the Introduction to Multilayer Networks Project, in which students are exposed to multilayer networks. The type of multilayer network that I am interested in captures each of the diverse types of relationships between the nodes into a separate layer of the network.

An example of such a network is the terrorist network in Fig. 1. I also provide the link to the comprehensive review (including temporal networks, networks of networks, and interdependent networks) which can be found in [16]. For visualization for multilayer networks, I suggest that they try Pymnet [44] found at http://people. maths.ox.ac.uk/kivela/mlnlibrary/, Muxviz [26] found at http://muxviz.net/, or Gephi [10] found at http://gephi.org/.

The students' assignment is to create a classroom multilayer network whose nodes are the students in the current class. Different student attributes get captured, which allow the formation of edges/relationships between the students, and are categorized into layers of a multilayer network. The students decide on the relationships they wish to capture, the end goal being to partition the class into teams of 3–4 students to write a research paper together. An expansion of this will follow in Section 5.2.

The second assignment for this period of time is to create or search for a network to analyze during the course of study, called the Network Profile Summary (Sect. 5.4). Each of these networks serves as data that each student analyzes for his/her homework by applying his/her understanding of the topics introduced in class that week. Thus the assignments are personalized, as the student chose the data, while all the students try all the learned concepts of the week and report their observations into one presentation slide per week (PowerPoint or L^AT_EX). The requirement of summarizing their findings in one slide provides the opportunity for the student to

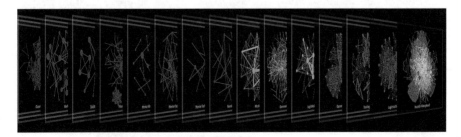

Fig. 1 An example of multiplex networks, in which each layer captures different relationships, such as friends, training, classmates, meetings, and operations

present synthesized information. Presenting the observed results enables the student to identify and explain the "why" behind the "what" of the findings, rather than asking the authority, the professor. Each student individually gives a 5-min presentation on his/her Network Profile Summary, based on the topics learned on that particular week. Presentations are followed by in-class discussion of that week's topics on a variety of networks that students present in class. An extended discussion on the Network Profile Summary follows in Sect. 5.4, and Gera et al. examine it in detail in [36].

Starting with the fourth week of the quarter is a transition to the following schedule:

- Mondays and Wednesdays: lectures.
- Tuesdays: teams meet to work on the Research Project (see Sect. 5.3). Each year a research topic is provided to the class, and each team finds a new methodology for solving one of the couple of choices of the open problems. While teams work on their projects, I work with each team to validate the direction of the research approach and to answer questions.
- Thursdays: students give their presentations on the Network Profile Summary (Sect. 5.4), followed by team discussions contrasting the results presented.

5 Student Learning and Assessment

The point value of the class activities are summarized in Table 1, with a longer discussion on each activity following the table.

5.1 In-Class Participation (70 Points)

The interactive teaching style requires everyone to participate in classroom discussions. Students are encouraged to be engaged in these discussions while giving everyone else a chance to confirm their understandings or mend their confusions. The in-class conversations allow students to modify and improve existing perceptions about the network science topics.

Table 1 The breakdown of points for the final grade

Activity	Points (of 300 total)
In-class participation	70
Introduction to Multilayer Networks Project	30
Network Profile Summary	100
Research Project	100

Assessments of learning: Participation is measured by evidence of class preparation, interactions during class. This is objectively measured by asking relevant questions, showing the ability to express critical thinking, and making connections even if they are not correct. These behaviors show whether students are actively engaged or passively listening.

5.2   Introduction to Multilayer Networks Project (30 Points)

The following has been used as an introductory project, modifying it for different cohorts of students, and it works well each time. While I talk about edge coloring in graph theory, there is a different purpose for the categories that form the colors there versus the layers; and thus this is the first exposure to multilayer networks.

As students will work in teams for the main Research Project detailed in Sect. 5.3, the first activity of this course provides the teams for the Research Project. That is, while learning about multilayer networks, students produce a multilayer network of their current class and identify a possible breakdown into teams to complete the Research Project. This way, while students start to think of multilayer networks for the first time, they have an interest in listening to the various solutions since (1) they thought about the problem as the whole class has the same task, and (2) they will be affected by the created teams.

Students work based on a description provided in advance and detailed below. Then they present the Introduction to Multilayer Network Project results at the end of the third week of classes. Each team has 10–13 minutes to convince me that their proposed teams (and reasoning for the team formation) should be the one to be adopted for the classroom. The following summarizes the project as given to the students:

Description: Research is a major component of this course. Since literature shows the best research is done in teams whose members have diverse backgrounds to integrate the research endeavor [40, 50], students are tasked to partition the class so that each team can accomplish the research goals. The students are provided with the class roster for MA4404 and some attributes that help them decide what relationships to add between the students. Such an example is shown in Table 2.

Goal: Students are asked to create a multilayer network (each attribute is captured in a different layer) and partition the MA4404 class into research teams. Ideally, members should complement each other based on the given attributes, and additional information can be collected. The goal is to minimize variability among performance of the teams, rather than to maximize the performance of one team.

Data: Table 2 presents sample data format used to create the class network, which could be augmented by other characteristics as the class sees useful. Each row represents a student, and each column captures the entry of that attribute per student.

Table 2 Possible data for the Introduction to Multilayer Networks Project

Name	Major (dual degree with)	Known coding language (first/second), Beginner (B) / Intermediate (I)/ Advanced (A)	Military service	Graduating month/year	Other relevant skill(s)	Previous partners
Student 1	MA	Python (B)	Army	June '17	Good speaker	Student 1
Student 2	CS	Python (A) and R (I)	Navy	Sep '17	Good writer	Student 4, 8
...	SE	None	Air force	Dec '17	Visual	...
Student n	OR	R (A)	Marines	Sep '17	Detailed	...

Tasks:

- Describe the methodology of network creation: Students must identify nodes, edges, and layers. Visualization is optional as they use their creativity to explain the network.
- Describe the methodology of team assignment: Students must describe the methodology for team creation, identify what characteristics were most important in selecting individuals for teams, and reasons why.
- Describe the results: Students must present their proposed teams and argument for why this distribution of talent meets the goals established for this project. Multiple solutions may be presented if needed, but not encouraged.
- Present conclusions and future work: Students expand on what they took away from this task. They provide other attributes they believe could inform these results and explain why those attributes matter.
- Assess learning: The assessment will be driven by the accuracy and creativity of the model and its solution. However, the following should shape the presentations and be taken into account: helpful visual aids and clear, complete, and organized presentation (labeled pictures/tables).

5.3   The Research Project (100 Points)

The goal is to explore a novel research topic and learn the process of turning an exploration into a research paper. To begin this process, each student will review one paper that will be used as one of the references for the project and turn in one to two paragraphs (or up to a page) synthesizing the paper. This allows students to critically look at the research, carefully analyzing examples of papers to follow as they work on their project.

I provide an environment in which students have a chance to think creatively and make educated hypotheses. As researchers learn a great deal from both success and failures, I maintain the perspective that mistakes are inevitable, and progress still happens since the missteps spark creativity and deepen understanding. While negative results are not always desirable, they will not impact the final grade if the procedure to obtain the hypothesis is correct. The final goals for the project are:

- A short, 5-minute update of the team research idea and findings each Tuesday.
- A final 20-minute presentation during the last week of classes.
- A team research paper (about 10 pages), due to the week of finals.

In 2017, the topic for the Research Project was to create a mathematical model for synthetic multilayer networks, as will be described next.

Creating multilayer synthetic networks (or generative models): We live in a connected world, where networks dominate our economy, our environment, and our society. Understanding these networks can aid researchers in devising plans for devastating natural disasters, such as the eruption of the Eyjafjallajökull volcano in 2010 [56] or the Ebola outbreak [42]. While real networks are insightful, they come with challenges: they are usually hard to obtain (such as repetitive samples of the same type of network) to create temporal networks; data collected to create networks may contain personally identifiable information, or the sampled data may be at the wrong scale; or it can be very time-consuming to create several data sets for analysis. Thus, researchers desire methodologies to create synthetic networks that mimic the real ones and that allow the researchers to change the parameters to create different scales of networks that have similar properties to those observed in the real networks.

Goal: The goal of this project is to create networks that are multilayer, have a varying parameter to get different scales, and have similar scaled properties to real ones at the layer and global levels (matching the properties of the real one when the synthetic network is at the same scale as the real network).

Data: For a multilayer data set, the following link for the European Union Airline Data can be used as an example: http://faculty.nps.edu/rgera/MA4404/EUAirports. zip. Students were encouraged to search other data sets as well. Larger data sets are available on Clauset's website [21] at https://icon.colorado.edu/.

A multilayer network has two or more layers based on the type of edges (See Fig. 1). A longer discussion can be found on Domenico's website [30] at https:// comunelab.fbk.eu/multinet.php or in Kivela et al.'s (2014) paper [43]. The choices for visualization tools are MuxViz [26], Pymnet [44], or Gephi [10] used to visualize each layer individually, or anything like it.

Assessments of learning: The weekly updates serve as formative assessments in preparation of the summative ones. I strongly suggest to students that they build their PowerPoint and research paper weekly, so I can provide feedback. I use the rubric in Table 3 for assessing the presentations. The outline below guides the Research Project that will emerge as a paper:

1. 25 points for ongoing weekly progress.
2. 75 points for research paper: Abstract, 5; Related Work, 10; Methodology, 10; Ingenuity (or reasoning for the existing method), 15; Analysis, 30; Conclusion, 5.

Table 3 Assessment rubric for slides and all the presentations, see Gera et al. Reference [36]

Criteria	Task	Detailed Step (0–10 points)
Content	An analysis is performed	Correct analysis synthesizing learned concepts
		A (9–10 points) Relevant and clearly explained findings, insightful contextualization of findings, and thoughtful synthesis and interpretation of metrics
		B (8 points) Minor errors
		C (7 points) Significant errors *D (6 points)* Major conceptual errors
		F (0–5 points) Little to no work of merit
Presentation	Results are presented	Clarity and style of graphics (could they be presented in a more significant way)
		A (9–10 points) *Slide Deck* Clear and succinct slides, correct spelling and mathematical notation, figures and tables are labeled and have captions consistent in tense and active voice, references are provided, thoughtful synthesis and interpretation of metrics *Conveying the information* Clear verbal explanation, correct use of terminology while explaining, clear and loud speaking
		B (8 points) Minor errors
		C (7 points) Significant errors *D (6 points)* Incoherent presentation
		F (0–5 points) Little to no work of merit

This paper's quality is given by knowledge integration:

1. Accuracy and vision: The modeling assumptions need to be appropriate, and the model needs to be checked against true network(s). The publication needs to give insight beyond a restatement of existing work and the exposition of the raw analysis of data; however, it should be related to existing work so that it has a place the current field of research.

2. Critical reasoning and exposition of relevant course material: Contrast the newly introduced methodology/parameter to existing ones. Present arguments for this methodology, and test statistics demonstrating competence with the content of complex networks. Explain connections to the real world and the observations/implications of the found results.

3. Clarity: Students must communicate the problem and questions addressing the introduced methodology and approach, their insights, solutions, and remaining open questions. Students are asked to make their explanations concise by eliminating unnecessary verbiage.
4. Rigor and precision: The resulting paper must be mathematically precise (using proofs if such results are presented) and logical in its reasoning throughout. Any methodology used should be justified, and limitations or assumptions should be clarified.

The two main parts of the project are:

- Theory development: Students propose a theoretical direction and present reasons for the new methodology.
- Data set analysis: students must compare the networks chosen for the Research Project to the synthetic networks they create, with the goal of showing similarities and discrepancies. Previous years' work on the Research Project has materialized in the following publications [5, 7, 11, 19, 22, 23, 25, 37].

5.4 Network Profile Summary (100 Points)

By the end of the third week of classes, each student must choose a network that (s) he is interested in understanding. For the remaining of the quarter, (s)he will be analyzing the chosen network and present results about it. This type of a project transitions from knowledge exposure to practice (without mimicking the instructor's particular example).

During and outside the class, students have the opportunity to experiment on their network, exploring the variety of topics as they present and as their intellectual curiosity inspires them. At the end of each week, each student creates one presentation slide (or two slides if absolutely needed) with the performed analysis of the topics presented that week, which is applied to his/her individual network.

Each Thursday four to five students present their findings, followed by a class discussion of all the networks' analysis (similarities and dissimilarities). Each student presents three times during the quarter, each being worth 20 points; the final presentation wraps up the takeaways from the study of the network and incorporates the personal updates provided weekly, and it is worth 40 points. The final presentation slides are due the last day of classes.

Data: Students search for data based on their interest. They can also bring the data they analyze for their theses, as they usually take this course during the time they work on the thesis projects (allowing students to incorporate network theory into their own research). I also provide scholarly sources and data that can be downloaded from a list compiled over the years http://faculty.nps.edu/rgera/MA4404/NetworkProfileSummaryResources.html. One can also collect personal data from Facebook or LinkedIn (anonymized and not published); also based on hashtags, one can easily collect data using Netlytic [38].

Examples: Now included are some examples from the 2017 cohort presentations, with the approval of the students. These slides have not been published; they are duplicates of the slides presented during the regular Thursday individual Network Profile Summary classroom presentations.

Major (Maj) Daniel Funk created the Global Maritime Transportation network (Figs. 2 and 3). There are 120 "ports and chokepoint," and the edges were built based on data of ports' exports and imports.

The data originated from CIA World Factbook https://www.cia.gov/library/publications/the-world-factbook [32]. The data is separated into a sea layer and a

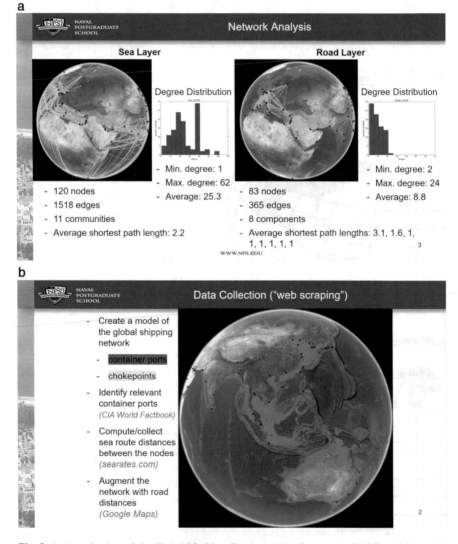

Fig. 2 An introduction of the Global Maritime Transportation System, by Maj Daniel Funk. (**a**) Highlighting the two types of nodes, (**b**) Highlighting the two layers

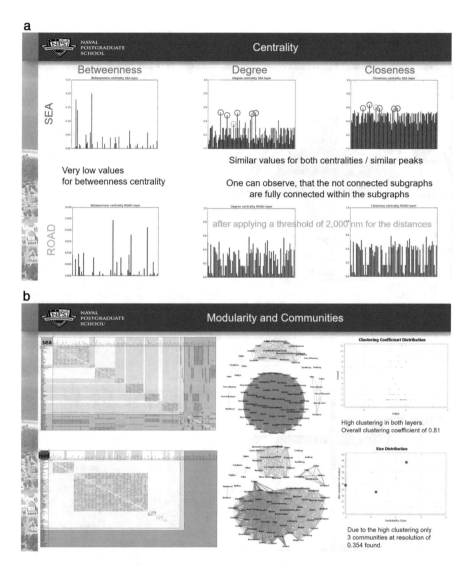

Fig. 3 The two layers of the Global Maritime Transportation System, by Maj Daniel Funk. (**a**) Centrality analysis, (**b**) Modularity and communities

road layer based on real travel distances (in nine nautical miles) on sea and land routes between the locations.

The data was collected from Bing Maps, Google Maps, and SeaRates https://www.Searates.com [4]. The PowerPoint slides present the results of modularity and community detection. Community detection partitions the network into groups by maximize modularity; it is one approach to studying communities in networks [49].

Captain (CPT) Brian Weaver analyzed the Storm of Swords data from Game of Thrones (Figs. 4 and 5). The data was collected and used in the article [12] to see who is the most degree central node.

a

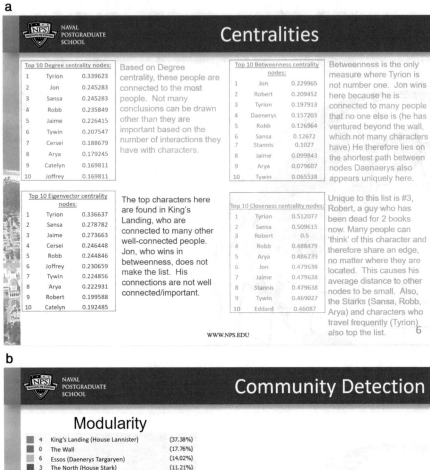

b

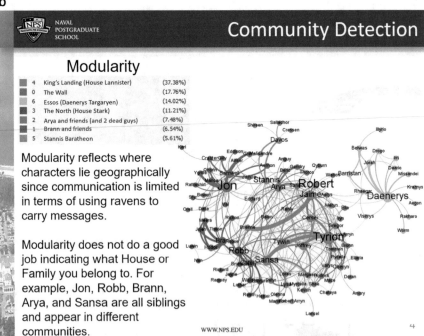

Fig. 4 The Storm of Swords of Game of Thrones network, by CPT Brian Weaver. (**a**) Centrality analysis, (**b**) Louvain community detection

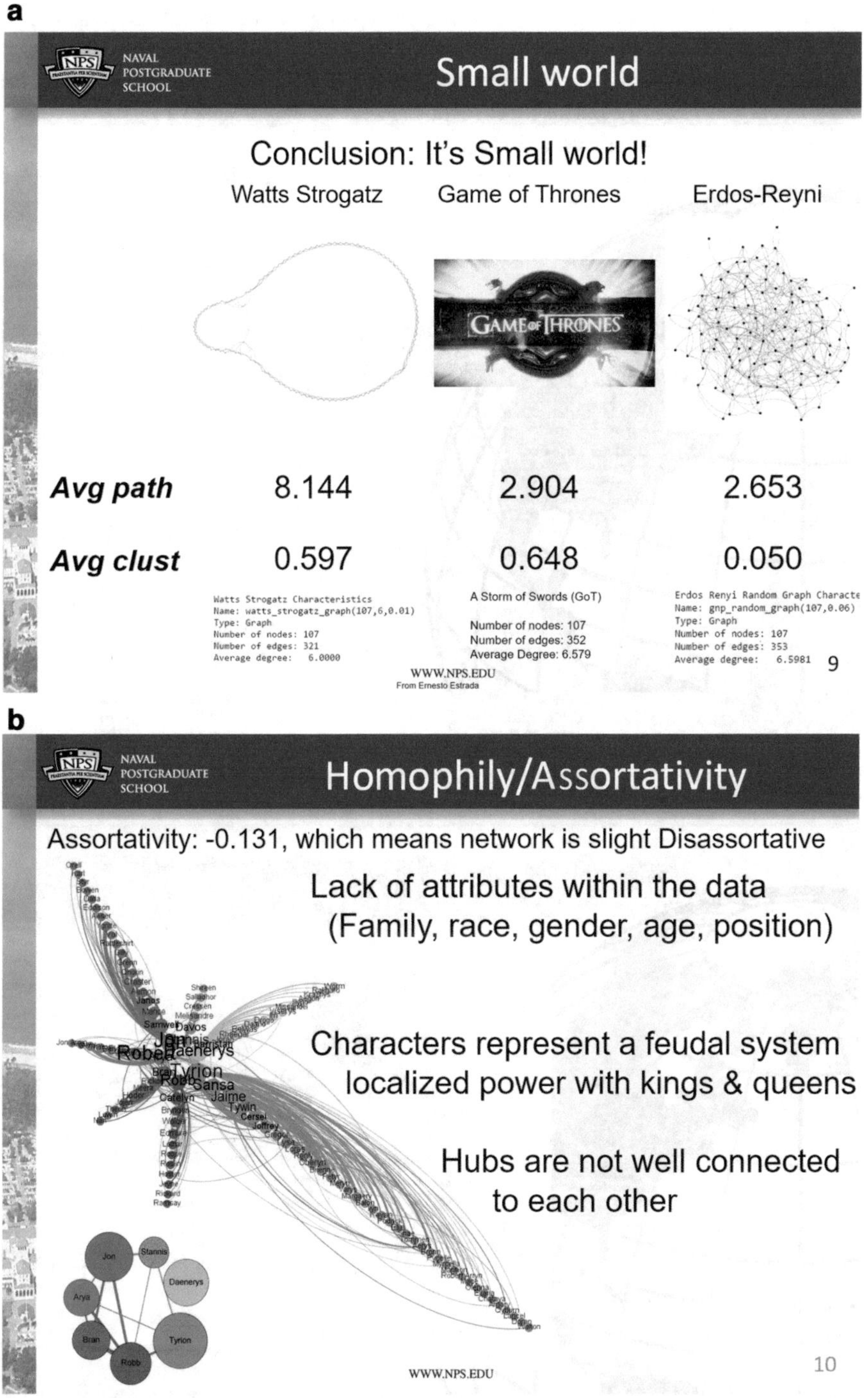

a

b

Fig. 5 The Storm of Swords as a small world and its homophily, by CPT Brian Weaver. (**a**) Comparing to Watts–Strogatz and Erdos–Renyi, (**b**) Homophily and assortativity analysis

Brian augmented their analysis based on the topics studied in the class. The examples particularly present the Louvain community detection results, in which the Louvain method is one way to partition the network in community based on maximizing modularity [14]. Porter et al. provide a comprehensive article on other methods of community detection [51].

Lieutenant Commander (LCDR) Kevin Garcia presented the Tesla Superchargers network (Figs. 6 and 7). The data and distances between the Superchargers were

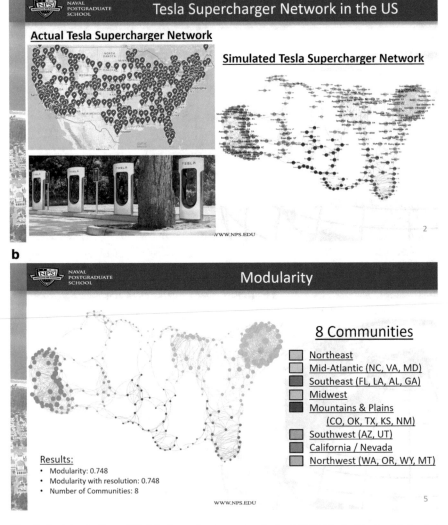

Fig. 6 An introduction of Tesla's Superchargers network, by LCDR Kevin Garcia. (**a**) A view based on the geolocation of the data, (**b**) Louvain community detection

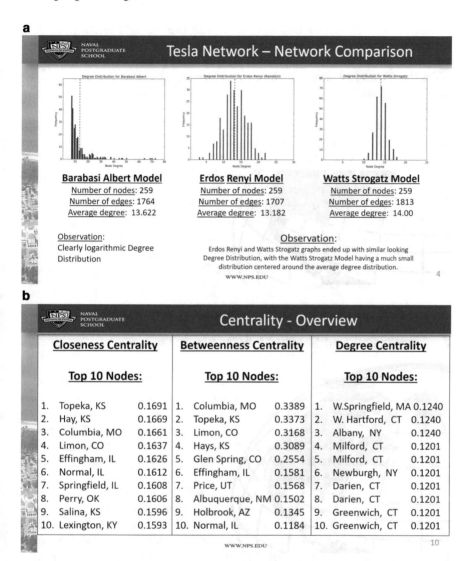

Fig. 7 Centralities and comparable synthetic networks, by LCDR Kevin Garcia. (**a**) Highlighting the different types of nodes, (**b**) Highlighting the different types of nodes

compiled by Maj Daniel Funk, using Tesla Supercharger locations obtained from www.Tesla.com.

The network has 259 nodes and 1700 edges based on the distance travelled on a full battery assumed to be 250 mile range. The actual range for different Tesla models varies from 240 miles up to 335 miles. Edges connect charging stations of distances up to 250 miles apart.

Kevin noted that, "While many networks slowly evolve over time, Tesla (and specifically Elon Musk) built the Tesla supercharger network because he was told that he wouldn't be able to. Instead of it evolving over time, he forced it, to make cross-country routes in a Tesla possible. This caused there to be very few nodes that are out on the periphery and not connected well to the graph. There are still several nodes that have a high degree in the metropolitan areas."

He concluded in his presentation slides [35] that for future work one should…. "analyze the network using all distances from each charging station to all other charging stations, not just the ones within 250 miles. From this information, edge weights could be added based on distances."

A more detailed explanation of the Network Profile Summary can be found in [36]. Some of previous students' work on the Network Profile Summary has become part of their theses publications [6, 34, 46, 54].

Assessment rubric for weekly presentations (and final presentation): This knowledge, skill, and ability assessment are both formative and summative assessment, allowing students to incorporate weekly feedback to refine the final project. A refined comprehensive presentation slide deck including the weekly draft presentation slide and a conclusion is due during the finals week. The details of the assessment are captured in Table 3.

6   Conclusion and Feedback from Students

This course and the teaching methods practiced have one underlying principle in the design: allow students the freedom in choosing what they analyze, while the instructor provides minimum guidance and explanations of the observed phenomena on the networks that students analyze. The instructor presents the high level overview of the topics covering the "why study" (during lectures and using TED talks) and "how to do it". However, the students apply the concepts to their chosen network, presenting the reasoning for what is observed.

Much like a building must pass certain control thresholds and a certain minimum quality control before moving on to the next stage, there is a need for a process approach to teaching that allows all students to obtain an accepted/approved level of understanding of newly learned topics before exposing them to additional information. With the current method, I believe that student's understanding is supported to be above a certain threshold by increasing their interest (letting them choose the network they study) and by asking them to find the "why" behind the "what is observed." I strongly believe that while I can provide them answers (i.e. teaching them), they only hear and understand it if they ask the questions to which I provide the answers. The practices described in this article allow the students to ask the questions, in order for them to hear my answers (the messages I express through my teaching) or find their own answers (which certainly empowers them and makes the inforamtion relevant and interesting).

The structure of the course facilitates an environment in which students can learn at their pace and depth. Each student is in control of the depth (s)he goes to understand her/his network Network Profile Summary, and each student must at a minimum apply all the topics (s)he was exposed to each week. The students then present their analyses and conclusions of what they learned of their network based on the new concepts learned that week. This step-by-step exploration of the unknown allows the students and professor to have an incremental approach to navigating through their network. Furthermore, this learning style is unlike the traditional one in which students mimic the examples worked in the book or classroom. This promotes creativity and allows the student to decide how to synthesize the information rather than mimicking an existing template.

Secondly, students work on Research Projects as a team, as they most probably work on everything else in the military environment. This way everyone contributes based on their strength(s), they see how useful they are to the team, and they build up confidence. The fact that the topic given is open for research rather than being a solved problem, allows them to consider ideas beyond what people in the field would normally think of, as well as freeing them from trying to come up with the answer that the professor already had. It allows them to experience what it means to do research. Neither the graduation nor the grade depend on the result itself but rather on creativity and the use of critical thinking based on the newly learned concepts.

The third reason is that pursuing their individual, or team's, interests while learning network science enables the student to correlate the learned concepts and methodologies to their preexisting knowledge. This is because they have freedom in their actions while taking responsibility for the choices of data, methodology, and presented results. These choices make the class relevant and allow students to synthesize the information they are exposed to, which makes the topics accessible in the future as need arises.

There are multiple advantages and ramifications to allowing the students to practice concepts on their personal network used for the Network Profile Summary since there are no expectations of certain results, and nobody to compare against as is traditionally done with homework. The expectation is that they make sense for the network, or the student finds an explanation if they don't make sense. Students first and foremost feel empowered by taking responsibility for their own learning since they have to understand and explain the results. Also, they answer questions over the interpretation of the results, as they are the experts of their own network. This motivates them to search for the reasons behind what is observed and gives them confidence in their findings.

Using the Network Profile Summary gives them the potential to obtain the needed understanding of the concepts before moving on. This is based on the observation that junior teachers get a deeper understanding of topics when they have to teach them to others, trying to find the best way to explain the topic and explaining the reasons behind the presented information. Lastly, it allows the students to further present a synthesis of the results obtained into convincing, coherent, and cohesive arguments for their personal network and team project.

Following is the feedback from students in the course:

"This class is structured in the way I thought graduate school classes should be structured. Open research questions were effective in inspiring more advanced learning. An excellent class."

"Great course! I enjoyed exploring the course concepts by implementing them on my network profile. Great way to learn!"

"This was an engaging and interactive course. It covered an incredibly interesting topic and the instructor did a great job bridging the 'math' with the real world applications. I enjoyed working in the team for projects and believe that this is by far the best way to learn. The lectures were interesting, improved my understanding of the material, and contributed directly to the quality of the projects. The instructor is passionate about the topics and passed that enthusiasm to the class. I wouldn't change anything."

"Best. Teacher. Ever. Loved the course, and if you had a mic Ralucca, this is the part where you would drop it and walk away."

Acknowledgments I would like to thank the reviewers for their helpful and constructive comments that greatly improved the quality of the current manuscript. This material represents the position of the author and does not reflect the official policy or position of the US Government.

References

1. Gephi. http://gephi.org/. [Online; accessed 2017-12-31].
2. Python. https://www.python.org/. [Online; accessed 2017-12-31].
3. R software. https://www.r-project.org/. [Online; accessed 2017-12-31].
4. Searates.com. https://www.SeaRates.com. [Online; accessed 2017-12-31].
5. Allain, R., Gera, R., Hall, R. and Mark Raffetto, (2016) M. Modeling network community evolution in YouTube comment posting. In International Conference on Social Computing, Behavioral-Cultural Modeling, & Prediction and Behavior Representation in Modeling and Simulation (BRIMS). http://sbp-brims.org/2016/proceedings/LB115.pdf.
6. Allen, G. (2017) Constructing and classifying email networks from raw forensic images. Master's Thesis, 2016.
7. Alonso, L., Crawford, B., Gera, R., House, J., Knuth, T., Méndez-Bermúdez,, J. and Miller, R. (2017) Identifying network structure similarity using spectral graph theory. Applied Network Science,
8. Barabási, A-L. (2016) Network Science. Cambridge University Press,
9. Barabási Albert, R.(1999) Emergence of scaling in random networks. science, 286(5439):509–512,.
10. Bastian, M, Heymann, S., Jacomy, M, et al. (2009) Gephi: an open source software for exploring and manipulating networks. Icwsm, 8:361–362,.
11. Max Berest, Ralucca Gera, Zachary Lukens, N Martinez, and Ben McCaleb. Predicting network evolution through temporal twitter snapshots for Paris attacks of 2015. In International Conference on Social Computing, Behavioral-Cultural Modeling, & Prediction and Behavior Representation in Modeling and Simulation (BRIMS). http://sbp-brims.org/2016/proceedings/LB111.pdf, 2016.
12. Andrew Beveridge and Jie Shan. Network of thrones. Math Horizons, 23(4):18–22, 2016.

13. John Biggs. What the student does: Teaching for enhanced learning. Higher education research & development, 18(1):57–75, 1999.
14. Vincent D Blondel, Jean-Loup Guillaume, Renaud Lambiotte, and Etienne Lefebvre. Fast unfold- ing of communities in large networks. Journal of statistical mechanics: theory and experiment, 2008(10):P10008, 2008.
15. Phyllis C Blumenfeld, Elliot Soloway, Ronald W Marx, Joseph S Krajcik, Mark Guzdial, and Annemarie Palincsar. Motivating project-based learning: Sustaining the doing, supporting the learning. Educational psychologist, 26(3–4):369–398, 1991.
16. Stefano Boccaletti, Ginestra Bianconi, Regino Criado, Charo I Del Genio, Jesus Gomez-Gardenes, Miguel Romance, Irene Sendina-Nadal, Zhen Wang, and Massimiliano Zanin. The structure and dynamics of multilayer networks. Physics Reports, 544(1):1–122, 2014.
17. Charles C Bonwell and James A Eison. Active Learning: Creating Excitement in the Classroom. 1991 ASHE-ERIC Higher Education Reports. ERIC, 1991.
18. Ann L Brown and Joseph C Campione. Guided discovery in a community of learners. The MIT Press, 1994.
19. Samuel Chen, Joyati Debnath, Ralucca Gera, Brian Greunke, Nicholas Sharpe, and Scott Warnke. Discovering community structure using network sampling. In 32nd ISCA International Conference on Computers and Their Applications. Springer, 2017.
20. Ken Cherven. Mastering Gephi network visualization. Packt Publishing Ltd, 2015.
21. Aaron Clauset. Index of complex networks.
22. Brian Crawford, Ralucca Gera, Jeffrey House, Thomas Knuth, and Ryan Miller. Graph structure similarity using spectral graph theory. In International Workshop on Complex Networks and their Applications, pages 209–221. Springer, 2016.
23. Brian Crawford, Ralucca Gera, Ryan Miller, and Bijesh Shrestha. Community evolution in multiplex layer aggregation. In Advances in Social Networks Analysis and Mining (ASONAM), 2016 IEEE/ACM, pages 1229–1237. IEEE, 2016.
24. Gabor Csardi and Tamas Nepusz. The igraph software package for complex network research. Inter- Journal, Complex Systems, 1695(5):1–9, 2006.
25. Benjamin Davis, Ralucca Gera, Gary Lazzaro, Bing Yong Lim, and Erik C Rye. The marginal benefit of monitor placement on networks. In Complex Networks VII, pages 93–104. Springer, 2016.
26. Manlio De Domenico, Mason A Porter, and Alex Arenas. Muxviz: a tool for multilayer analysis and visualization of networks. Journal of Complex Networks, 3(2):159–176, 2015.
27. Ton De Jong and Wouter R Van Joolingen. Scientific discovery learning with computer simulations of conceptual domains. Review of educational research, 68(2):179–201, 1998.
28. Edward L Deci, Robert J Vallerand, Luc G Pelletier, and Richard M Ryan. Motivation and education: The self-determination perspective. Educational psychologist, 26(3–4):325–346, 1991.
29. Josep Diaz, Mathew D Penrose, Jordi Petit, and Maria Serna. Linear orderings of random geometric graphs. In Graph Theoretic Concepts in Computer Science, volume 1665, pages 291–302. Springer, 1999.
30. Manilo De Domenico. Multilayer networks.
31. Leonhard Euler. Leonhard euler and the konigsberg bridges. Scientific American, 189(1):66–70, 1953.
32. CIA Factbook. The world factbook. See also: https://www.cia.gov/library/publications/the-world-factbook, 2010.
33. Scott Freeman, Sarah L Eddy, Miles McDonough, Michelle K Smith, Nnadozie Okoroafor, Hannah Jordt, and Mary Pat Wenderoth. Active learning increases student performance in science, engineering, and mathematics. Proceedings of the National Academy of Sciences, 111(23):8410–8415, 2014.
34. Daniel Funk. Analysis of the global maritime transportation system and its resilience. Master's Thesis, 2017.
35. Kevin Garcia. The tesla's superchargers network. unpublished.

36. Ralucca Gera, Jessica M Libertini, Jonathan W Roginski, and Anthony Zupancic. The network profile summary: exploring network science through the lens of student motivation. Journal of Complex Networks, 2017.
37. Ralucca Gera, Ryan Miller, Miguel Miranda Lopez, Akrati Saxena, and Scott Warnke. Three is the answer: combining relationships to analyze multilayered terrorist networks. Advances in Social Networks Analysis and Mining (ASONAM), 2017.
38. A Gruzd. Netlytic: Software for automated text and social network analysis, 2016.
39. Aric Hagberg, Pieter Swart, and Daniel S Chult. Exploring network structure, dynamics, and function using networkx. Technical report, Los Alamos National Laboratory (LANL), 2008.
40. Martin Hoegl and Hans Georg Gemuenden. Teamwork quality and the success of innovative projects: A theoretical concept and empirical evidence. Organization science, 12(4):435–449, 2001.
41. Paul A Kirschner, John Sweller, and Richard E Clark. Why minimal guidance during instruction does not work: An analysis of the failure of constructivist, discovery, problem-based, experiential, and inquiry-based teaching. Educational psychologist, 41(2):75–86, 2006.
42. Maria A Kiskowski. A three-scale network model for the early growth dynamics of 2014 west Africa ebola epidemic. PLoS currents, 6, 2014.
43. Mikko Kivela, Alex Arenas, Marc Barthelemy, James P Gleeson, Yamir Moreno, and Mason A Porter. Multilayer networks. Journal of complex networks, 2(3):203–271, 2014.
44. Mikko Kivela. Multilayer networks library for python (pymnet). http://people.maths.ox.ac.uk/kivela/mln_library/. [Online; accessed 2017-12-31].
45. Eric D Kolaczyk and Gabor Csardi. Statistical analysis of network data with R, volume 65. Springer, 2014.
46. Ryan Miller. Purpose-driven communities in multiplex networks: Thresholding user-engaged layer aggregation. Master's Thesis, 2016.
47. Michael Molloy and Bruce Reed. A critical point for random graphs with a given degree sequence. Random structures & algorithms, 6(2–3):161–180, 1995.
48. Mark EJ Newman. The structure and function of complex networks. SIAM review, 45(2):167–256, 2003.
49. Mark EJ Newman. Modularity and community structure in networks. Proceedings of the national academy of sciences, 103(23):8577–8582, 2006.
50. Max Nielsen-Pincus, Wayde Cameron Morse, Jo Ellen Force, and JD Wulfhorst. Bridges and barriers to developing and conducting interdisciplinary graduate-student team research. Ecology & Society, 2007.
51. Mason A Porter, Jukka-Pekka Onnela, and Peter J Mucha. Communities in networks. Notices of the AMS, 56(9):1082–1097, 2009.
52. Dale H Schunk. Self-efficacy and academic motivation. Educational psychologist, 26(3–4):207–231, 1991.
53. Jeffrey Travers and Stanley Milgram. The small world problem. Psychology Today, 1:61–67, 1967.
54. Scott Warnke. Partial information community detection in a multilayered network. Master's Thesis, 2016.
55. Duncan J Watts and Steven H Strogatz. Collective dynamics of 'small-world' networks. nature, 393(6684):440–442, 1998.
56. Dunn S. Wilkinson, S.M. and S. Ma. The vulnerability of the European air traffic network to spatial hazards. Natural Hazards, 60(3):1027–1036, 2012.

Advances in Nontechnical Network Literacy: Lessons Learned in Tertiary Education

Paul van der Cingel

1   Introduction

Real-world problems, challenges, developments, and systems are displaying increasing levels of complexity [1, 4]. Complex issues require studying the interplay between agents and factors [2]. Therefore, analytical methods should be counterbalanced with methods that stress synthesizing skills on the systemic level. In education, this could pose serious challenges to curricula, in particular in programs that are using real-world contexts while teaching mainly analytical skills to their students. Much of our existing education, it seems, still shows a considerable legacy of centuries of traditional reductionistic sciences. The persistence of this path dependency can be easily seen when studying the twenty first century skills that are currently being promoted as a cornerstone for future education. A recent report from the World Economic Forum shows analytical skills being mentioned explicitly [3]. However, systemic awareness, sense of interconnectivity, or synthesizing skills remain unmentioned.

In Sect. 2, we will first shed some light on the concept of real-world complexity. Section 3 will focus on the context of tertiary professional education in the Netherlands. In Sect. 4, we describe the method of nontechnical network literacy that was used. Section 5 captures lessons learned from various experiments at Windesheim University of Applied Sciences. This chapter concludes with questions that set the agenda for future experiments in Sect. 6.

P. van der Cingel (✉)
Windesheim University of Applied Sciences, Zwolle, Netherlands

© Springer International Publishing AG, part of Springer Nature 2018   45
C. B. Cramer et al. (eds.), *Network Science In Education*,
https://doi.org/10.1007/978-3-319-77237-0_3

2   Real-World Complexity

Even though complexity science has been around for more than three decades, its multidisciplinary nature makes it hard to reach consensus about the mere definition of complexity [4, 5]. In order to move forward in creating a methodology of nontechnical network literacy, we chose to define complexity by taking the perspective of etymology. As "plexus" is Latin for "wicker work," complexity should always be about interconnected or interdependent agents, organizations, and things. Furthermore, we chose to label an issue as "complex" only if its interconnectedness was ambiguous and unstable over time. Thus, we made our definition applicable to many real-world problems and challenges. The choice to focus on ambiguity and instability also enabled us to separate complex problems from simple and complicated problems [6].

Thus, our working definition of complexity consisted of three elements:

1. *Connectivity*: complex issues always involve various kinds of agents and factors that are influencing each other.
2. *Ambiguity*: there are multiple ways in which they can influence each other.
3. *Instability*: the way they influence each other can change over time. Using this working definition, Table 1 shows some examples of real-world complexity in the Netherlands.

In many programs in Dutch tertiary (or higher) education, these types of complex issues are commonly used, not only as case studies but also as starting points for assignments. Issue A, for example, is almost certainly discussed in applied health science programs but also in applied information technology programs and management programs. It is to this higher professional education that we turn now.

3   Higher Professional Education in the Netherlands

Dutch higher education has two distinct paths. The first path offers scientific research-oriented education at universities, embedded in a bachelor-master structure. Currently, about 260,000 students take this path at some 20 institutes. The

Table 1   Real-world complexity: three examples from the Netherlands

A. Some 390 local governments are responsible for the provision of healthcare services for their citizens. They work with suppliers from a cluster of 6000 healthcare organizations, resulting in 1-year contracts or longer term commitments. Healthcare services are categorized into 120,000 different product codes

B. Students map all known existing initiatives concerned with improving youth health in a city. They come up with 180 initiatives, many of which are interrelated in a variety of ways. For instance, they found financial relationships between initiatives but also personal connections and connections of legal ownership

C. A 15-year-old girl posts a birthday party invitation on Facebook. She not only invites her 78 friends but also ticks the box "friends of friends," which causes her invitation to go viral. Ultimately, thousands of youths turn up in the small village of Haren in September 2012, and the resulting riots came to be known as Project X Haren [7]

Table 2 Examples of Bachelor programs at Windesheim University

Domain	Examples of professional fields
Technology	Engineering, Industrial Design, IT
Media	Journalism
Health	Applied Gerontology, Speech Therapy
Business	Financial Services Management, Business and Information Management
Social Work	Socio-Pedagogical Care, Child Care Management
Education	Teaching in primary and secondary education, subjects ranging from Sports to Theology
Multidisciplinary honors programs	Global Project and Change Management

second path offers higher professional education at universities of applied sciences, embedded in a bachelor structure. Currently, over 440,000 students follow this track at 37 institutes [8].

Our experiments took place in the context of higher professional education. Specifically, we tried out network literacy education at Windesheim University of Applied Sciences. This university has been ranked as one of the Dutch top institutions for a number of years. With a student population of over 25,000 and a staff of 2000, it hosts 50 bachelor programs [9], preparing students for a great variety of professions. Table 2 provides examples of typical 4-year bachelor programs from various domains at Windesheim University.

In the old days, students encountered real-world issues mainly during professional internships. During the rest of the curriculum, real-world problems only surfaced when mentioned by lecturers or in the literature. This situation of "selective exposure" has changed. Nowadays, external professionals and other stakeholders are increasingly invited to present their complex issues to students in various parts of curricula. Many times, the issues are turned into assignments or research questions that students have to tackle, e.g., various innovative design processes and consultancy assignments. Item B in Table 1 provides an example.

Within the context as presented here, we now turn to question how network literacy experiments were conducted in various programs at the university. In order to answer this question, we first have to describe the methodology that we developed.

4   A Method of Nontechnical Network Literacy

In this section, our method is described by its basic assumptions, key characteristics, and learning objectives.

Assumptions Setting up experiments is often preceded by a formal assessment of the initial conditions. In our case, we chose not to conduct such formal research, in order to initiate the experiments as quickly as possible. Instead, we operated pragmatically, conducting experiments where we were given the opportunity, evaluating learning along the way. The analogy with a "fail forward and pivot" strategy seems fitting. We framed all experiments in a setting of "innovative frontrunners" with possible ripple effects to other parts of programs.

The nontechnical network literacy concept we used was based on three assumptions. First, as pointed out in the introduction and in Sect. 1, real-world issues that students have to tackle are characterized by increasing complexity. They display growing connectivity, ambiguity, and instability. Second, as mentioned in the introduction, much of the existing education emphasizes analytical skills, thus teaching students to take a reductionist approach in tackling real-world problems. This makes students focus on aspects of the problems at hand, ignoring interconnectedness between the various aspects, let alone ambiguity and instability. Third, visualizing the interconnectedness between agents and factors in real-world issues can help students to balance the analytical approach by zooming out to the system level. This should strengthen their synthesizing skills. Network visualizations are very well suited to display interconnectedness. As Manuel Lima, author of *Visual Complexity*, states: "... the network is a truly ubiquitous structure present in most natural and artificial systems you can think of, from power grids to proteins, the internet and the brain. ... Networks are an inherent fabric of life." [10]

4.1 Characteristics of Preparation and Learning Objectives

The experiments were not designed as a stand-alone series of sessions. Instead, we developed plug-in workshops, because we consciously wanted to operate within existing curricula that were based on problem-centered education. This ensured that there already was a group of students who were working on a real-world complex issue. Obviously, this accelerated the speed of operations. On the other hand, it implied that our first step was to go out and find teachers that were willing to open up their lessons to the workshops.

The next step was to negotiate timing and quantity, in relation to expected outcomes. What was the best moment for the workshop to be plugged in? For optimal efficacy, we expected students to have familiarized themselves with the complex issue at hand on a basic level. That meant they would have had some weeks to do desk research, interviews, etc., in order to gather information on agents and factors of their complex issue. The following questions were: how much time was the teacher willing to free up?, and how many workshops could be conducted in that time frame? Given that setting, the last question was: what outcomes could reasonably be expected?

We related outcomes to the following *learning objectives*:

1. The workshop should create systemic awareness and strengthen students' synthesizing skills. By introducing key network science concepts like clustering and hubs in a nontechnical way, students' insights into the problem should deepen.

2. The workshop should enhance the quality of students' research into the problem. The network diagram should be used to map out the known knowns as well as the known unknowns, thus enabling students to prioritize their research agenda.
3. The workshop should foster dialogue within student teams or between students and real-world stakeholders. This, in turn, should enhance the emergence of a shared vision on the problem.

Before each experiment, we sat down with the teacher to discuss the outcomes that could be expected given the scope of the actual setting. This also was the best moment to ask the teacher for a list of the complex issues that students were working on. This ensured that the workshops could be fitted optimally into the existing lessons.

We ended up doing a variety of experiments. The smallest ones were one-session workshops of less than 3 hours, and the largest ones were three-session workshops of over 3 hours each. Table 3 provides an overview.

4.2 Characteristics of the Workshops

Taking a problem-based approach implied that students came to the workshop with a known real-world complex issue, which they ideally had been researching for some time. Many students were at the point of experiencing lack of coherence and high levels of fragmentation due to information overload.

In each workshop, whether small or large, we tried to pay attention to five elements. Table 4 shows these workshop elements and relates them to expected outcomes and also to the three learning objectives.

Table 3 Overview of experimental network literacy workshops at Windesheim University

Domain	Bachelor program	Workshop aspects	Frequency
Media	Journalism	3rd-year students; one-session workshop	Done twice (April 2016 and September 2016)
Health	Applied Gerontology	2nd-year students; two-session workshop	Done once (January 2017)
Business	Business & Information Management	3rd-year students; one-session workshop	Done twice (November 2015 and November 2016)
Education	Teaching in primary and secondary education	1st-year students; two-session workshop	Done twice (both May 2017)
Multidisciplinary honors programs	Global Project and Change Management	2nd-year students; three-session workshop	Done once (May 2017)

Table 4 Experimental workshop elements related to outcomes and learning objectives

Element number	Description	Expected outcome and *[corresponding learning objective]*
1	Creating systemic awareness	Basic knowledge of concepts of complexity and systems thinking. Applying this to the complex issue at hand [1]
2	Connecting the dots	Hand-drawn network diagram of agents, factors, and their interconnectedness of the complex issue at hand [1]
3	Creating dialogue	Deepened insight into the complex issue at hand. In student teams, fostering a shared vision on the complex issue [1, 3]
4	Improving the network diagram by integrating dialogue outcomes	Deepened insight into the complex issue at hand. In student teams, fostering a shared vision on the complex issue [1, 3]
5	Studying the network diagrams	Applying basic knowledge of network science to the complex issue at hand. Mapping research, creating insight into known knowns and known unknowns [1, 2]

Table 5 A taxonomy of problems based on systems thinking

	System components: agents and factors	Type of connectivity
Simple problems	Few	Unambiguous and stable
Complicated problems	Many	Unambiguous and stable
Complex problems	Few or many	Ambiguous and unstable

Each workshop ended with what might be called a network topology of the real-world complex issues that students were working on. We explicitly told students that this status quo was very likely going to change in the future, because of the inherent ambiguous and unstable nature of complex connectivity. In a few experiments, we had time to address this topic, which we called network dynamics. We asked students to develop what-if scenarios, which they could start by adding any change in either a network component or a connection. In the following description of workshop characteristics, we will focus on network topology as we had very limited experiences with network dynamics.

The first part of the workshop was all about the creation of systemic awareness. We did this by introducing the concepts of complexity and systems thinking and relating those to the students' problems, using the taxonomy shown in Table 5.

This led most students to discover that they were dealing with a complex problem because of ambiguity and instability in the connections between agents and factors. We then discussed the consequences of this discovery, focusing on the pitfalls of over-analyzing the systems' components while forgetting to pay attention to the effects of their interconnectedness. In this way, students were nudged into thinking about the follow-up question: how can we ensure that we pay explicit attention to connectivity? At this stage, we presented network literacy as an answer to that question and told them it was time for the second part of the workshop: "connecting the dots."

The "connecting the dots" assignment instructions were as follows:

- Hand-draw a network diagram of the interconnected agents and factors in your complex issue.
- If you are working in a team, please make this sketch individually. Later, you will discuss individual diagrams and create a final diagram as a team.
- The goal of this exercise is not to get a complete picture, merely to map out all you know from your research.
- Use different shapes and different colors to denote categories in agents and factors.
- Mark different kinds of connections, relationships, or links. For example, you could draw thicker lines to denote strong connections and thinner lines for weaker ones.
- There is no such thing as "the perfect diagram," so feel free to take another sheet of paper and start anew.
- If you are struggling, try the two-step approach: first map out agents and factors and then think about connections.
- Figure 1 illustrates this two-step approach.

This exercise usually took classically trained students out of their analytical-reductionist mode, as they had to explicitly think about not only the dots but also about the connections between them.

When these sketches had been drawn, we started the third part of the workshop: creating a dialogue. The assignment looked like this:

- Share your network diagram with another student. If you work in a team, share with a team member.
- Sharing means presenting your diagram to the other student and, reversely, listening to her presentation.
- After listening to the presentation, ask each other questions about the diagrams. Your questions should not only address system components but also connections. When possible, ask questions about components or connections that you think are missing in the diagram.

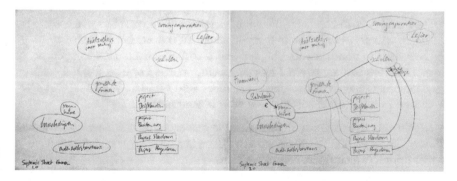

Fig. 1 A two-step approach to create a hand-drawn network diagram

This assignment usually led to productive dialogue between students. Within teams, it opened up the possibility of a more coherent shared vision on the problem at hand.

The fourth part of the workshop consisted of integrating insights from the dialogue assignment into the network diagrams. We provided students with flip chart paper and asked them to revise their original sketch, taking into account the outcomes of the dialogue. Teams could take this opportunity to integrate their individual diagrams into one "team diagram." Thus, this part of the workshop ended with a collection of network diagrams, hand-drawn on flip chart paper. We asked students to hang them on walls or windows in the workshop location. If time permitted, we then did another round of peer review, asking students to study all diagrams and leave at least one comment using sticky notes.

The fifth part of the workshop was to study these network visualizations using very basic network science insights. Depending on how much time we had to conduct the workshop, we discussed questions such as:

- Does your diagram show clustering? If so, what can you say about the degree of homogeneity/diversity within the clusters? Do you consider this an opportunity or a threat to your complex issue?
- Do you see a difference between strongly connected dots (the core) and weakly connected dots (the periphery)?
- Do you see hubs? Who can explain why they are relevant for fast flows (of information, money, etc.) between, e.g., core and periphery?
- Are clusters connected? Who can explain why cluster connectors are relevant for fast flows (of information, money, etc.) between, e.g., core and periphery?
- How many different types of relationships did you draw?
- We are now "zoomed out," looking at all diagrams on the wall from a distance. Do you spot missing links in your diagram? If so, what makes them potentially interesting?

This part of the workshop was also when we returned to the research that students had done prior to the workshop. This research had given them basic information about their complex issue, enabling them to draw agents, factors, and connections. If time permitted, we gave them the following research-plotting assignment:

- You have been researching your complex issue over the past few weeks. Take the list of references or literature that you compiled and number all items.
- Now turn to your network diagram and plot the numbers near components or connections that are covered by that specific source. For instance, if a database containing school alliances has the number 4 on your list, you could plot a "4" near the cluster of schools that you drew in your network.

Evaluating this exercise, we explained to students that they now had a map of "known knowns" and "known unknowns." The last category would enable them to better prioritize their future research efforts.

5   Lessons Learned

First and foremost, we learned that nontechnical network literacy showed great potential to empower students to deal more effectively with real-world complex issues. Thus, it proved to be of considerable value to Dutch higher professional education. By making and studying a topology of their complex issue with hand-drawn network diagrams, students were being nudged into consciously strengthening their synthesizing skills. Many times, we saw that this was the first moment in their school career that they were urged to take the systems' perspective and were being explicitly encouraged to extend their thinking beyond "the dots," taking into account the influence of "the connections." One student expressed her empowerment by stating:

> To me, the workshop was a real eye-opener. I am not very good at relating and integrating concepts and research findings. Thus, my research mostly results in lots of incoherent lists of findings. By connecting the dots, I gained far more insight.

As stated earlier, most experiments covered network topology but did not leave us enough time to cover network dynamics. That leaves us with future lessons to be learned about teaching students to better deal with the uncertainty of instability in their complex problems or challenges.

From an organizational perspective, we can conclude that within the course of a school year, we were able to experiment with nontechnical network literacy by conducting plug-in workshops in existing curricula in professional higher education. Moreover, a basis for continuation emerged, as all workshops were evaluated more or less positively. First, students were asked to rate the value of the workshop by reviewing the three learning objectives. The majority of students reacted positively. Second, teachers were asked to review the outcomes and were also asked if they were willing to host the workshop again next year. Nearly all teachers answered positively.

Experience offered other insights, such as get to know what is already going on and who is involved.

1. The existence of programs or parts of programs that offer problem-centered education. In higher professional education, this almost automatically implies that you have found real-world issues.
2. The existence of teachers who are open to listening to your case for network literacy and are consequently willing to free up time for an experimental workshop. By reaching out to the teachers, you are bound to find out whether the real-world issues at hand display complexity. In later years of 4-year bachelor programs, they often are.

By sitting down with the teachers to prepare for the workshops, we learned a second lesson: try to get the expectations right. It helps to show the teacher some slides you are planning to show in your workshop and monitor her/his reaction. Sometimes, teachers implicitly expect you to help students to "solve" the complex problems or challenges. If so, you can point out that accepting complexity means letting go the assumption that one can solve the issue like a complicated puzzle.

It may also help to explicitly discuss the three learning objectives, and ask the teacher where he/she thinks the students' learning curve will be the steepest. We learned the added value of this point the hard way. One of the collaborating teachers wanted to plug in the workshop in an early stage of the curriculum. During the preparation, we didn't bring up the topic of research. By the time we were conducting the workshop, we soon discovered that the students hadn't spent proper time to do research into their complex issue. Consequently, they did not know which agents and factors were involved, and they weren't able to make a purposeful network diagram. We improvised by asking students to draw a conceptual network diagram as a starting point for research.

It may also help to ask the teacher to be present during the workshop. This may prove helpful for two reasons. First, it simply shows students that the teacher is really committed to the workshop and its added value to the program. Second, it provides the teacher with an opportunity to learn about network literacy, perhaps silently adjusting pre-existing assumptions and judgments and even become a "Network Literacy Ambassador" within the institution.

In conducting the workshops, we learned many straightforward lessons. Here we name four:

1. Use the native language as much as possible. During the first part of the workshops (Creating awareness; see Table 4), students were taken out of the dominant analytical problem-solving mode into a more systemic mode, requiring synthesizing skills. This meant that many students were experiencing a large gap between their regular way of working and a new systemic way of working. Using English terms and concepts, such as emergence, nonlinearity, nodes, and edges, enlarged the gap and dampened the learning curve. It seems safe to state that using plug-in network literacy workshops in existing curricula is best combined with using the native language. This proved to be an important lesson, albeit an unwelcome one, because it urged the need for learning materials in Dutch. Apart from a Dutch version of the network literacy "Essential Concepts and Core Ideas" brochure [11], suitable learning materials are still very scarce.
2. Stress the importance of connections explicitly and more than once. During the dialogue assignment, we learned to explicitly urge students to talk about the connections in their diagrams. It also proved necessary to explicitly evaluate this part of the assignment afterward. When we failed to do so, we frequently saw students discussing only the components in their diagrams with each other, forgetting the connections in the networks.
3. Use larger paper and smaller pens. In the course of time, we found ourselves switching from A4/letter size paper to A3/ledger/tabloid size paper and replaced the whiteboard markers with a wide range of color pens with regular tips. Even in workshop stage 4 (Table 3), when using flip chart paper, we encouraged students to use the regular pens, having discovered that many students quickly filled the smaller paper with their sketch. This problem recurred when students were transferring their sketches to the flip chart paper and used whiteboard markers.

4. Hold onto the flip chart diagrams. In workshop settings with multiple sessions, we learned that holding on to the flip chart papers yourself was the best way to go. Most of the times, the flip chart papers weren't allowed to stay on the walls of the room. Letting students take their paper home with them, guaranteed that some of them forgot to take it with them for the next session. Letting them take a photo instead proved much more effective.

6   Questions for Future Experiments

1. Network visualizations work reasonably well as a means for dealing with complex issues. Nevertheless, we encountered people who seemed to be almost allergic to (network) diagrams and reluctant to draw dots and lines on a piece of paper. What other methods are available to use in network literacy education?
2. Plug-in workshops are an effective way to get started quickly. How can we scale up? Should we use the lessons learned to create a stand-alone, complete "complexity curriculum"? Or should we connect the various experiments across the university, in order to create a "learning community of network literacy Ambassadors"?
3. How can we create a lasting effect from the workshops? On several occasions, students and teachers were happy with the workshop, and some great network diagrams were made. However, chances were most of these outcomes were not used in the weeks and months after the workshop.
4. How can software help us in workshops? Should we stop drawing on paper and let students draw on screens with stylus pens? Could this also open up possibilities to convert hand-drawn diagrams into more stylized diagrams, enabling us to get network statistics? Could it also open up possibilities to simulate what if-scenarios and study network dynamics?

Acknowledgments   The author thanks Prof. Dr. Lasse Gerrits, Chair, Political Science, leading the Governance of Innovative and Complex Technological Systems, University of Bamberg, Germany, for helpful comments on an earlier version of this chapter. All remaining errors are the author's responsibility.

References

1. Gosselin, D. and Tindemans, B (2016) *Thinking Futures – Strategy at the edge of Complexity and Uncertainty.* Leuven: Lannoo Publishers. 65.
2. Heydari, B., & Wade, J. (2014). *Complexity: Definition and Reduction Techniques, Some Simple Thoughts on Complex Systems.* Kerala, India: Centre for Studies in Discrete Mathematics.
3. Colander, D. and Kupers, R. (2014) *Complexity and the art of public policy -Solving society's problems from the bottom up.* Princeton & Oxford: Princeton University Press.
4. Chen, C. & Crilly, N. (2016) Describing complex design practices with a cross-domain framework: learning from Synthetic Biology and Swarm Robotics. Research in Engineering and Design, 27. Dordrecht: Springer. 291. doi:https://doi.org/10.1007/s00163-016-0219-2

5. Chic-Chun Chen (2016) *From modularity to emergence: a primer on the design and science of complex systems*. DOI: https://doi.org/10.17863/CAM.4503. Report number: CUED/C-EDC/TR.166, Cambridge: University of Cambridge.
6. Project X Haren. (2017) https://en.wikipedia.org/wiki/Project_X_Haren Accessed 1/19/18.
7. Dutch Higher Education. (2018) https://www.studyinholland.nl/education-system. Accessed 1/19/18
8. Windesheim University of Applied Studies. https://www.windesheiminternational.nl/ Accessed 1/19/18
9. Manuel Lima, M. (2013) *Visual complexity –Mapping patterns of information*. New York: Princeton Architectural Press.
10. World Economic Forum (2016) New Vision for Education: Fostering Social and Emotional Learning Through Technology. Report. https://www.weforum.org/reports/new-vision-for-education-fostering-social-and-emotional-learning-through-technology. Accessed, 1/19/18.
11. Network Literacy: Essential Concepts and Core Ideas (2015) https://sites.google.com/a/binghamton.edu/netscied/teaching-learning/network-concepts

Part II
Creating New Degree Programs

Network Science Undergraduate Minor: Building a Foundation

Chris Arney

1 Introduction

As an interdiscipline, network science (NS) encompasses several intellectual perspectives and subjects, including mathematics, sociology, psychology, information science, and biology. Unlike traditional statistical approaches, which assume data independence, NS assumes dependent interconnection that can then be modeled by a graphical structure. Through its methodology, NS focuses on the impact of the relationships among individual entities. This enables NS students to identify and analyze the essential characteristics and properties of the structures and processes on the network. Many systems rely on both physical connections, such as those supporting communications or shared computing, and sociocultural connections, such as the web of trust-based relationships that exist among people [1, 2]. NS methods provide the tools to study many elements of contemporary information society as described in Easley and Kleinberg [3]. NS is an active, growing discipline in the modern information world, and consequently, undergraduate institutions are beginning to offer academic courses and programs in NS. This paper will discuss and provide information on the NS education minor program at USMA as an example of the type of program that can educate undergraduates in this important area of modern science.

C. Arney (✉)
United States Military Academy, West Point, New York, NY, USA

© Springer International Publishing AG, part of Springer Nature 2018   59
C. B. Cramer et al. (eds.), *Network Science In Education*,
https://doi.org/10.1007/978-3-319-77237-0_4

2 Background

Network science has been enhanced by an entirely new arsenal of methods and models. As an interdiscipline, its methods connect quantitative concepts from several disciplines such as mathematics, biology, physics, computer science, data science, operations research, cognitive science, sociology, and information science, with qualitative approaches from sociology, linguistics, social science, behavioral science, and psychology. By its very nature, NS uses methods and theories that involve emerging areas of science such as complex adaptive systems, cooperative game theory, agent-based modeling, and data analytics. NS is modeling with a structured, entity-linked, framework. Much of the strength of network modeling is in its ability to embrace the complexities of the real world. NS has become an important and empowering form of interdisciplinary analysis and problem solving to confront the important issues of the society. Network science research is still building its underlying framework, inventing new computational tools and techniques, and organizing network relevant data. This makes NS an exciting, relevant, modern field of study at all levels of education.

3 Network and Information Science

As the world becomes more globally connected, it is increasingly defined by networks. Our ability to quantify underlying factors that drive these networks continues to improve [ibid] and [4]. Whether systems are made up of physical, technological, informational, or social networks, they share many common organizing principles and thus can be studied with similar approaches. With this powerful framework, we have discovered how networks form, grow, transform, dissolve, evolve, and behave; how they facilitate the flow and the spread of information, behaviors, resources, and diseases; how knowledge transforms the network and how the networks transform knowledge; what mechanisms drive network formulation and operation; and how we can intervene to disrupt or rehabilitate networks. Undergraduate education is designed for the future success of the students, and reports on the future of network science given by National Research Council Network Science Committee [5] and Coronges et al. [6] provide these application domains with great potential for impact by network science:

- Group decision-making
- Personal and population health
- Biological systems and brain activity
- Socio-technical infrastructure
- Human-machine partnerships

The academic minor program that we have established at USMA seeks to build a foundation in the basic science of networks for our students. These foundational elements include:

- Mathematical computation to measure properties and modeling methods for dynamic processes
- Data analysis and processing methods for data visualization, network inference techniques, and tools to navigate and synthesize network data
- Theory and mechanisms for understanding network processes, such as diffusion, control, and coordination
- Ethics in collecting, storing, sharing, and using information

One way to define a network is to establish its components (nodes, links, data, processes, etc.), its properties (dynamic, functional, layered, etc.), and its applications (logistics, flow, transportation, information transfer, metabolic networks, social networks, organizational networks, etc.). The foundational research management report on network science written by the National Research Council Network Science Committee [5] used a layered approach for the network roles – physical, communicative, informational, biological, and social/cognitive. Modern networks often contain a critical information layer that can be exploited; security is always an issue. Therefore, cybersecurity is a big part of informational network science [7].

Many networks are too complex to rely on visualization to achieve understanding. Defining, computing, and measuring well-defined properties can counter those misguided visual perceptions and improve network modeling and analysis. According to Arney [8], the roles of the discipline include: working with information scientists to build explicative (based on a theory or conjecture) and empirical (based on statistical analysis of data) models, creating appropriate measures for important applications, finding appropriate properties, formalizing measurement systems for properties, and increasing security of the information in the network system.

It is interesting to look at a topical family tree for network science. While the historical development and the contribution relationships are subjective (based on an aspect of the transition but possibly not all elements) and incomplete (often limited to one primary contributor being listed when there were many), one can see in Fig. 1 the disciplinary web that brings together some of the many elements that have contributed to network science.

4 Undergraduate Curricula

At USMA, we have seen that undergraduate students have considerable interest in network science. Courses and programs that traditionally were offered at the graduate level are attracting undergraduates' interest. As social media and online systems are seen to be parts of NS, student interest in NS has grown in leaps and bounds. NS is a valuable addition to the academic program at USMA for several important reasons.

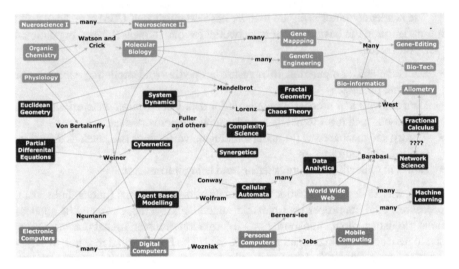

Fig. 1 Disciplinary connection network model showing links between subjects (Green denotes biological subject, maroon denotes mathematics subjects, and orange denotes computing subjects). (Fig. 1 was assembled by network science student John McCormick as part of his senior thesis at USMA)

4.1 Network Science Minor

At USMA, the network science minor consists of five academic courses:

- One required course (Foundations of Network Science). This course provides the basic introductory framework necessary for learning and understanding network science.
- Three electives within the minor. Students are required to select three courses from an extensive, flexible, multidisciplinary list of possibilities, thus designing their program to complement their other coursework and interests. The selected courses must fulfill the required elements of theory, modeling, and application. The theory requirement can be fulfilled by a course such as:

  - Linear Algebra
  - Discrete Mathematics
  - Graph Theory
  - Sociological Theory
  - Design And Analysis Of Algorithms

- The modeling requirement can be fulfilled by:

  - Human Physiology
  - Genetics
  - Data Structures
  - System Simulation

- Mathematical Modeling
- Deterministic Models
- Stochastic Models
- Systems Dynamics Simulation

• The application elective has many possible courses, some that fulfill this requirement are:

- Infrastructure Engineering
- Computer Networks
- Artificial Intelligence
- Cyber Security Engineering
- Comparable Military Systems
- Supply Chain Engineering
- Urban Geography
- Geographic Information Systems
- Cyber Policy, Strategy, and Operation
- Leading teams
- Human-Computer Interaction
- Econometrics
- Network Science for Cyber Operations

• One required course in integration of their learning. This final course in the network science minor is a capstone of their learning within the minor. Students select a topic of interest in conjunction with a faculty advisor. This integration experience is an opportunity for cadets to synthesize previous material in tackling a demanding problem. This course provides experiences in presenting technical results both orally and in a technical report. Recent titles of projects that fulfilled this requirement include "Modeling Household Vulnerability in Kampala, Uganda," "Optimizing Intelligent Walks for Dark Network Discovery," "A Heuristic for Approximating a Minimum Dominating Set of an Arbitrary Directed Graph," "Undersea Communication Cables," and "Analyzing a Network of NATO Scientists."

5 Network Science Minor Learning Model

Academic curriculum has three principal structural components: breadth, depth, and integration. Network science, as an interdisciplinary minor, provides opportunity for additional study in an area beyond the major, thus reinforcing all three academic components. The network science minor also offers military-related knowledge, which enhances graduates' ability to meet the USMA overarching academic program goal to respond effectively to the uncertainties of a changing technological, social, political, and economic world.

- *Structure of learning experiences*: the structure of the network science minor program is designed to expose cadets to an emerging interdisciplinary field with enough depth to develop their modeling, analysis, and problem-solving skills. The program nurtures creativity, critical thinking, and self-directed learning through activities performed in theory, modeling, and application settings. The minor includes both the acquisition of a body of knowledge and the development of thought processes judged essential to understanding fundamental ideas and principles in network science.
- *Process of learning experiences*: the introductory course provides theoretical and mathematical foundations for further study and experience in applying network concepts to real-world problems. The three elective courses selected by the cadet give depth in the discipline. The integration course is generally taken last and provides an opportunity to work on a more complex problem involving analysis of real-world data or understanding of an important issue. The goal of the integration course is to enhance conceptual understanding of key concepts and to unify concepts through their application to a real-world problem. Learning takes place through various means such as reading, discussion, laboratory discovery, research projects, and leveraging the various emerging technologies in network science.

6 Fundamentals in Network Science Course[1]

This required course starts the educational minor by exposing students to the basic concepts of networks and gives them an opportunity to apply techniques learned in the course to real-world problems. Students develop skills and problem-solving strategies for modeling complex networks associated with physical, informational, and social phenomena. The Organizational Risk Analysis (ORA) software package (http://www.casos.cs.cmu.edu/projects/ora/) is used to investigate application problems and augment understanding of the course material. The course seeks to build the students' ability to understand and learn and solve problems through these following goals:

- *Acquire a base of knowledge*. Learn basic graph theory terms and acquire the fundamental skills and concepts required to conduct network analysis. These skills include representing networks of nodes and links with graphs and matrices, calculating network level and node level measures, identifying group structures, analyzing the distribution of centrality measures, and evaluating the underlying causes for link formation. Students learn to calculate and measure centralities (degree, closeness, betweenness, eigenvector), PageRank, homophily, transitivity, reciprocity, and balance.
- *Apply technology*. Use software to build and analyze models using network analysis software.

---

[1] Much of the new content for the Fundamentals of Network Science course was developed by Donald Koban in 2016 and 2017.

- *Communicate effectively.* The students improve their communication skills, both verbally and in writing by presenting article summaries throughout the course, writing a technical paper and creating a research poster.
- *Solve problems confidently and competently.* The students analyze problems by formulating models, discussing critical assumptions, analyzing networks, and interpreting the results.
- *Develop habits of mind.* Students practice several important intellectual elements such as creativity, critical thinking, curiosity, and teamwork.
- *Think with an interdisciplinary perspective.* This course embraces the complexity of modeling the structures and processes of organizations, systems, and networks [9].

The course textbooks are Newman [1] (2010), which provides an overview and introduction to network science, and Easley, Kleinberg [3] (2010), which provides more exposure to diverse applications. The videos watched in class include:

- http://www.wimp.com/wolvesrivers/
- https://www.youtube.com/watch?v=nJmGrNdJ5Gw
- http://www.opte.org/the-internet/
- http://www.ted.com/talks/nicholas_christakis_the_hidden_influence_of_social_networks

The data sets used include:

- Stanford Large Network Dataset Collection: http://snap.stanford.edu/data/
- Clauset et al. Collection [10]: http://tuvalu.santafe.edu/~aaronc/powerlaws/data.htm
- UC Irvine Network Data Repository: http://networkdata.ics.uci.edu/
- Mark Newman Collection: http://www-personal.umich.edu/~mejn/netdata/
- Sean Everton Noordin Top Terrorist Network: https://sites.google.com/site/sfeverton18/research/appendix-1
- Visualization of A Day in the Life of Americans: http://flowingdata.com/2015/12/15/a-day-in-the-life-of-americans/

The articles used for readings are an important component of the course. Getting the students into the network science literature early in the sequence of the program helps them in their courses later in the program. In addition to the two textbooks, the students also read articles [11–24].

7 Network Science for Cyber Operations[2]

This course fulfills the application component of the minor. The course goal is to enable students to confront cyberspace issues by modeling, solving, analyzing, and understanding problems involving cybersecurity on networks. Students learn about

---

[2]The content of the Network Science for Cyber Operation was developed by Professors Chris Arney and Natalie Vanatta, who have team-taught this course several times.

networks, perform complex modeling, work on a project of their choosing, write and present a book review, produce a poster on their project, and give presentations on their project. Much of the material is learned through individual reading, class discussions, and lectures by cyber professionals. The course material is supplemented by the many examples in the news related to cybersecurity. The course embraces the complexity of determining human-based security measures (utility functions) for the structures and processes of cyber systems and information networks. From a cyber perspective, this course explores the role of cyber policies and practices on networked systems, communities, and all of society.

Defensive cyber operations seek to maintain network performance and reliability while protecting and securing information and communication validity. Offensive cyber operations seek to disrupt network performance or damage the information and communications of an adversary. Cyber concerns are critical to the functionality of the Internet and the World Wide Web. The layers of the various networks within the dark or deep web can affect critical infrastructure and information used for national and global commerce and communications. Disruptions in the Internet or World Wide Web or lapses in their security put the infrastructure and economic health of the United States at risk. The sheer volume and heterogeneity of the data, the dynamics of evolving threats, and the sensitivity of normal and malicious behavior all pose major challenges. The course content includes:

- *Cryptography*: This material traces the history of cryptography from Caesar to modern elliptic curves. Students will explore the innovation of encryption through the mathematical foundations that they are built on. Emphasis will be placed on the military applications of encryption and how they have changed the face of the battlefield.
- *Cyber Mission Forces*: The Cyber Mission Forces are a new addition to the US military to concentrate on cyber-related military operations.
- *Internet of Things*: The Internet of Things is quickly becoming our new reality– or perhaps the greatest threat to our way of life. Students explore the complex issues (both technical and non-technical) that create the foundation for the Internet of Things.
- *Social Sciences*: There are many important aspects to cyber operations that are nontechnical in nature. Students study cyber policy, regulation, and law as it pertains to both military and nonmilitary operations. Special emphasis is placed on cyber ethics and the tug-of-war between privacy and security.
- *Data Analytics and Science*: Visualization and how to do data analytics for the security of large-scale networks.

The course uses *Network Science* by Albert-László Barabási [25] for its network science textbook along with each student selecting one of these three books as a second textbook:

1. *Cyber War* [26] (policy, regulation, and nation state actors)
2. *Code Book* [27] (historical to modern ciphers and a brief chapter on the future)
3. *Cybersecurity and Cyberwar* [28] (how the Internet and basic tenants in cyberspace work)

The students are asked to consider the course material in the context of the cyber book they select and these questions:

- What kind of public policy should be in place to regulate/manage access to information technology?
- Should we establish a list of laws and rules that carry severe penalties for violators, or should we use incentives, and build an infrastructure that guides usage?
- How can we identify trends found in social media to understand how changes in networks can signal shifts in beliefs and behaviors?
- How can social media be used to sway public opinion and create consensus to vote or protest?
- Who has knowledge about the security of the Internet (including the resource providers and resource consumers)?
- What is the complexity or interconnectedness of public systems networks?
- What kinds of policies or controls would help regulate security in such a complex system?
- What are the roles of the public and private sector in these public needs?

8 Network Science Problems in the Interdisciplinary Modeling Contest

The Interdisciplinary Contest in Modeling (ICM) http://www.comap.com/undergraduate/contests/index.html is a large international modeling contest conducted annually involving over 24,000 undergraduates from over ten different countries. For 4 days, teams of up to three undergraduates research, model, analyze, solve, write, and submit 20-page solutions to an open-ended problem. The solution papers are judged and categorized as described in the contest's history and overview [29]. The problems are chosen from three areas (network science, environmental science, or policy). Some of the popular network science problems from the contest are as follows:

2012 This problem required teams to investigate the relationships of the members of a criminal conspiracy network within a business organization through social network analysis of their message traffic. It required teams to understand concepts from the informational and social sciences to build effective network and statistical models to analyze the message data between 83 people involving over 400 messages that were categorized into 15 topics. In order to accomplish their tasks, the students had to consider many difficult and complex disciplinary and interdisciplinary issues. An Outstanding-category solution paper by Guo et al. [30] can be found in the *UMAP Journal*.

2013 This problem required teams to investigate the relationships of local and regional ecosystems to the global health of the planet. It required teams to understand concepts from the informational, environmental, and social sciences to build network and statistical models to track the potential changes in Earth's global health. In order to accomplish their tasks, the students had to consider many difficult and complex disciplinary and interdisciplinary issues. An Outstanding-category solution paper by Moitinho de Almeida [31] can be found in the *UMAP Journal*.

2014   The NS requirement investigated the relationships involved in network models for determining influence in a large coauthor network (Mathematician Paul Erdös' 511 coauthors) and measuring impact within a set of foundational papers within the discipline of network science. This problem required teams to mine a large data set and understand concepts from the informational sciences to build effective models for these complex phenomena. The problem contained many multifaceted issues to be analyzed and had several challenging requirements for innovative scientific and network modeling and analysis. In addition to network modeling, informational analysis, and data collection, the teams had to explain the nature of influence and impact in an academic social network and show how their models could be used to help make informed decisions. An Outstanding-category solution paper by Wang et al. [32] can be found in the *UMAP Journal*.

2015   The network problem for 2015 involved modeling human capital issues (especially employee churn) in a hypothetical organization of 370 employees with the intent of aiding managers and decision-makers to build successful systems for recruiting, hiring, training, and evaluating employees. Having teams analyze the network-related issues of human capital is a relevant issue in improving performance and profits of many modern organizations. An Outstanding-category solution paper by Blanc et al. [33] can be found in the *UMAP Journal*.

2016   The network problem was set in a historical context where society's information networks of five time periods (1870s, 1920s, 1970s, 1990s, and 2010s) were compared. By using the news and media networks of each period, measures for the flow of information relative to the value of information were established and compared. Teams used historical data and developed measures and models to determine what qualifies as news and to track the evolution of news throughout the ages. An Outstanding-category solution paper by Norman et al. [34] can be found in the *UMAP Journal*.

The ICM offers an opportunity each year for teams of undergraduate and high school students to tackle challenging, real-world problems that require skills in network science. ICM problems are open-ended, challenging problems. The complex nature of the ICM problems and the short time limit require effective communication and coordination of effort among team members. One of the most challenging issues for the team is how to best organize and collaborate to use each team member's skills and talents. Teams that solve this organizational challenge often submit excellent solutions. We have included information on the network science problems of the ICM because it can be a significant motivator for students to study more NS in their undergraduate programs.

9 Conclusion

The discipline of network science in the undergraduate curriculum has followed similar paths as operations research and computer science in the later part of the twentieth century [7, 35, 36]. If that trend continues, many colleges will have undergraduate network science programs similar to the one at the USMA described here. This the way that science and its curricula evolve and impact our society and education system.

References

1. Newman M E J (2010) Networks: an introduction. Oxford University Press, London.
2. Barabási A-L (2002) Linked: how everything is connected to everything else and what it means for business, science, and everyday life. Perseus, New York.
3. Easley D, Kleinberg J (2010) Networks, crowds, and markets: reasoning about a highly connected world. Cambridge University Press, Cambridge, UK.
4. Kivelä M, Arenas A, Barthelemy M, Gleeson J, Moreno Y, Porter M (2014) Multilayer networks. J Complex Networks 2(3): 203--271.
5. National Research Council Network Science Committee (2005) Network science. National Academy Press, Washington, DC.
6. Coronges K, Barabási A-L, Vespignani A (2017) Future directions of network science: a workshop report on the emerging science of networks. Virginia Tech Applied Research Corporation.
7. Arney D, Pulleyblank W, Coronges K (2014) Integrating information sciences: operations research, computer science, computational linguistics, analytics, network science, computational sociology, and applied mathematics. Phalanx 46(4): 29–35.
8. Arney D. (2012) Network science: what's math got to do with it? UMAP Journal 33(3): 185–191.
9. Weaver W (1948) Science and complexity. American Scientist 36: 536.
10. Clauset A, Shalizi C, Newman M E J (2009) Power-law distributions in empirical data. SIAM Review 51(4): 661–703.
11. Morris M, Kurth A, Hamilton D, Moody J, Wakefield S, The Network Modeling Group (2010) Concurrent partnerships and HIV prevalence disparities by race: linking science and public health practice. American Journal of Public Health 99(6): 1023–1031.
12. Macdonald B, Shakarian P, Howard N, Geoffrey M (2012) Spreaders in the network SIR model: an empirical study. arXiv:1208.4269v2.
13. Haugen K (2004) The performance-enhancing drug game. Journal of Sports Economics 5(1): 67–86.
14. Antal T, Krapivsky P, Redner S (2006) Social balance on networks: the dynamics of friendship and enmity. Physics D 224: 130.
15. Milgram S, Bickman L, Berkowitz L (1969) Note on the drawing power of crowds of different size. Journal of Personality and Social Psychology 13(2): 79–82.
16. Newman M E J (2001) Scientific collaboration networks: network construction and fundamental results. Phys. Rev. E 64: 16131.
17. Price D (1965) Networks of scientific papers. Science 149: 510–515.
18. Girvan M, Newman M E J (2002) Community structure in social and biological networks. Proc. Natl. Acad. Sci. USA 99(12): 7821–7826.
19. Christakis N, Fowler J (2007) The spread of obesity in a large social network over 32 years. New England Journal of Medicine 357(4): 3700–3709.

20. Mastrandrea R, Fournet J, Barrat A (2015) Contact patterns in a high school: a comparison between data collected using wearable sensors, contact diaries, and friendship surveys. *PLOS* 1.
21. Watts D, Strogatz S (1998) Collective dynamics of 'small-world' networks. Nature 393: 440–442.
22. Davis G, Yoo M, Baker W (2003) The small world of the American corporate elite, 1982–2001. Strateg. Organ. 1: 301–326.
23. Koban D (2015) A static random Bernoulli model for the analysis of covert networks. Military Operations Research Journal 20(4): 39–47.
24. Borgatti S (2005) Centrality measures and network flow. Soc. Networks 27: 55–71.
25. Barabási A-L (2013) Network science. http://Barabási.com/networksciencebook/
26. Clarke R, Knake R (2010) *Cyber war.* Harper-Collins, New York.
27. Singh S (2000) *The code book.* Fourth Estate, New York.
28. Singer P, Friedman A (2014) Cybersecurity and cyberwar. Oxford University Press, London.
29. Arney C, Campbell P (eds) (2014) Interdisciplinary contest in modeling: culturing interdisciplinary problem solving. COMAP, Bedford, MA.
30. Guo F, Su J, Gao J (2012) Finding conspirators in the network via machine learning. UMAP J 33(3): 275–292.
31. Moitinho de Almeida D, Shapiro E, Knight A (2013) Saving the green with the greens. UMAP J 34(2–3): 211–228.
32. Wang C, Gong M, Li Z (2014) Who are the 20%. UMAP J 35(2–3): 229–248.
33. Blanc G, Bristol H, Wang J (2015) Organizational churn: a roll of the dice. UMAP J 36(2): 113–136.
34. Norman A, Wyatt M, Flamino J (2016) Characterizing information importance and its spread. UMAP J 37(2): 121–144.
35. Arney C, Coronges K, Gera R, Roginski J, Sheetz L (2015) DoD's role in network science: NetSci symposia on models, teamwork, and education. Phalanx 48(1): 22–26.
36. Alderson D (2008) Catching the 'network science' bug: insight and opportunity for the operations researcher. Operations Research 56(5): 1047–1065.

Evaluation of the First US PhD Program in Network Science: Developing Twenty-First-Century Thinkers to Meet the Challenges of a Globalized Society

Evelyn Panagakou, Mark Giannini, David Lazer, Alessandro Vespignani, and Kathryn Coronges

1 Introduction

1.1 Purpose of the Program

As the world becomes more globally connected, it is increasingly defined by networks. Our ability to quantify underlying principles that drive network dynamics and evolution has vastly improved in the last decade. With roots in physical, information, and social sciences, network science provides a formal set of methods, tools, and theories to describe, prescribe, and predict network dynamics. Despite this formalization, there is still considerable debate over what constitutes the fundamental techniques, methods, and theories of network science. That is, how do we identify ourselves in a field of study that is by its very nature transdisciplinary and that has become so pervasive and is being taught across so many disciplines? Our PhD program attempts to establish, from the multitude of disciplinary methods and theory, a general framework that defines network science as a coherent field and will define the next generation of network scientists.

Dramatic improvements in information technology over the past 20 years, including increased storage capacity and computing power, have made it possible to archive and study multiple levels and multiple dimensions of biological and sociotechnical systems with unprecedented detail in areas such as communication, transportation, natural resources, infectious diseases, and political and cyber systems. Science and technology are growing exponentially, not only in terms of ideas and knowledge produced and spread but also in terms of the emerging applications that utilize this knowledge. The approaches and methods of network sci-

E. Panagakou (✉) · M. Giannini · D. Lazer · A. Vespignani · K. Coronges
Network Science Institute, Northeastern University, Boston, MA, USA

© Springer International Publishing AG, part of Springer Nature 2018   71
C. B. Cramer et al. (eds.), *Network Science In Education*,
https://doi.org/10.1007/978-3-319-77237-0_5

ence have synergies across disciplines and pervade data science and data analytic methodologies. The latter are the core expertise sought after by many companies in their hiring plans. In the 2011 McKinsey Global Institute (MGI) study [1], it was predicted that by 2018 the United States alone could face a shortage of 140,000 to 190,000 people with this expertise. In a more recent analysis, the 2014 MGI study [2] suggests that understanding networked capabilities – both technically and behaviorally – will transform organizational practices and will drive two-thirds of the value creation opportunities afforded by social technologies. Further, a series of reports from the National Science Foundation, National Institutes of Health, National Research Council, and Institute of Medicine [3] have highlighted two fundamental directions for future scientific progress, complexity and transdisciplinarity, both of which are hallmarks of network science. Research on network connections among multiple types and levels of "actors" offers a powerful paradigm to understand the workings of complex systems across broad areas of science, including information and technology, biological systems, health and health care, local and global political and economic processes, and sociotechnical infrastructures and sustainability.

Adopting a "network view" requires novel evaluations of, reconfigurations in, and innovations for standard methods of theorizing, data collection, and analysis. However, the new techniques capable of evaluating, designing, and influencing these systems and their interdependencies developed in network science and other computational methodologies have not yet been systematically formalized into an educational curriculum. We do not yet have a workforce with the background to effectively capitalize on these new techniques. To meet new challenges arising from an increasingly interconnected globalized society, we have developed a doctoral program to grow a new kind of scholars with an interdisciplinary quantitative and social scientific training, tailored to leverage these new capabilities.

2 Program Description and Overview

The Northeastern University's Network Science PhD program couples fundamental network science methods with disciplinary knowledge, enabling theoretical and substantive understanding of the appropriate use of network scientific tools and techniques. Thus, the PhD program is built on the following standards:

- Rigorous training in mathematical, computational, and theoretical concepts, fundamental to network science. Students gain this knowledge in the first 2 years of coursework with the core classes.
- Exposure to key tools and techniques of network modeling across disciplines – including data collection strategies, computing languages, modeling approaches, and theoretical and problem-solving strategies. While we cannot expect trainees (or even faculty) to be expert, or even proficient, in these tools, the understanding

of and respect for their potential contributions to novel interdisciplinary approaches to network science is paramount. Students gain this expertise through considerable hands-on core coursework, as well as through selection of elective courses that allow them to become proficient in a set of capabilities appropriate to their interests and research area.

- Experience with applied knowledge in disciplinary fields of study and the ability to frame the major problems in these fields into a set of network-based problems. Students gain this knowledge primarily through mentored research, starting in their second year. Applicants to the program select one of four concentration areas: natural sciences (physics, biology, ecology); social sciences; health science (epidemiology); and computer and information sciences. Tracks loosely guide the student through their coursework and function to help students navigate the selection of electives and dissertation advisor.
- Foundational training in all aspects of network science (e.g., approaches, languages, problems), beginning in the first year of graduate training, as it is necessary in order to build an inherently interdisciplinary science and the next generation of researchers and projects.
- Deep dive and practice in matching theoretical and substantive questions with the appropriate use of network-based tools and techniques. This knowledge is the most complex and nuanced and the hardest to assess. Students gain these perspectives from weekly journal club discussions, in which mathematical, conceptual, and even philosophical notions of the field are explored. In addition, active speaker and workshop series offer students a great breadth of scientific excellence across fields.

Finally, the program relies on a team of research-active faculty members to provide mentorship and advising.

The interdisciplinary nature of the proposed program draws students who are interested in applying network science in different areas and disciplines. The program has attracted students who are not only technically strong but are also interested in major scientific challenges and the solution of real-world problems. The program provides a path for students to acquire experience and skills in networks while at the same time being knowledgeable in a specific application area.

2.1 Program Structure

The doctoral program is an interdisciplinary program, supported by multiple colleges and departments at Northeastern University. The Network Science Institute at Northeastern University serves as the primary research organization for doctoral students. The Institute is currently made up of ten core faculty collocated in a modern, largely open floor setup. Our core faculty and affiliated faculty members are from a range of academic departments, including physics, political science, communication, computer science, health sciences, and business. Applicants to the doctoral program are evaluated on their academic readiness (e.g., GPA, exam scores), interest and understanding of network science (assessed by their personal statement), and, in some cases, research experience (while this is not required, exposure and success in research environments is weighed heavily). Successful applicants must show outstanding academic and intellectual capabilities as well as having interest in an area that is aligned with the Institute's projects. When students apply to the program, they must identify a focus area – political science, health science, computer science, or physics. We use these categorizations to help us create diverse cohorts and identify possible faculty mentors. Students invited into the program are offered fellowships for their first year, allowing them time to explore different areas of research and mentorship styles. By the end of the first semester of year 2, students will have selected a dissertation advisor from the core network science faculty or associated faculty.

2.2   Program Objectives

The purpose of the program is to build competence in network science through: (1) coursework, (2) research collaboration and exchange (through an extensive schedule of speakers, research visitors, workshops, and team meetings), and (3) independent (mentored) research on a range of large-scale projects that draw extensively from the multiple disciplines and tools offered by the field.

The key goals for the network science program are to:

- Develop interdisciplinary scientists who understand and appreciate the full scope of network science and who are poised to engage challenging questions across multiple disciplines
- Understand the diverse languages, foci, and tools of network research by introducing students early in their graduate training to both disciplinary and interdisciplinary orientations regarding the influence of interconnections in complex social, virtual, physical, and natural systems
- Acquire skills using relevant theories and advanced modeling methods
- Understand the range and value of different (qualitative and quantitative) data collection methods and analytic techniques
- Develop intellectual flexibility regarding approaches to network-based research
- Serve as a catalyst for a new generation of network research and the emerging field of network science
- Understand important issues in scientific careers, including challenges and opportunities (e.g., from the logistics of funding/publication to considerations of ethical, institutional, and societal challenges in scientific work)

We expect that the graduates from our PhD program will have acquired:

- Comprehension of the mathematics of networks, and their applications to biology, sociology, technology, and other fields, and their use in the research of real complex systems in nature and human-made systems
- Adequate knowledge on network modeling, on network data mining clustering, visualization techniques, statistical descriptors of networks and computational statistics, data acquisition and handling, measurement, and research design
- Ability to communicate network science concepts, processes, and results effectively, both verbally and in writing
- Preparation to enter many potential career paths including industrial research positions, government consulting positions, and postdoctoral or junior faculty positions in academic institutions

2.3   Admission Criteria and Process

Application materials include transcript(s), personal statement, three letters of reference, and the general GRE scores. Students are accepted with a bachelor's or higher degree in any field and should have either academic or work experience demonstrating a commitment to working in network science. Interest in the program has steadily increased with 13 applicants in 2014, 61 applicants in 2015, 71 applicants 2016, and 86 in 2017. Currently, the Institute is training 21 network science PhD students. Over the four admission cycles that have taken place, approximately 17% of the applicants have received an offer of admission. Successful applicants typically have an average undergraduate GPA of 3.61; verbal and quantitative GRE general scores at 85th and 86th percentile or higher, respectively; analytical GRE general score of 4.4; and a minimum TOEFL score of 100 (in the case of international applicants). Offers of admission are made based on the applicant's qualifications, the alignment of research goals with existing faculty, and space within the program. The students will obtain a PhD Degree in network science.

3   Characteristics of the PhD program

The PhD curriculum is designed to provide students with graduate-level understanding of foundational network science concepts. In addition to course evaluations, there are three assessments over the doctoral training: Qualifying Exam, Comprehensive Evaluation, and Dissertation Defense. The successful student will master the following fundamental skills:

- Comprehension of the mathematics of networks and their applications to biology, sociology, technology, and other fields.
- Statistical descriptors and biases of network data
- Measures and metrics of networks

- Network clustering techniques
- Network modeling
- Network data mining techniques from real-world datasets to networks
- Understanding process modeling on networks
- Network visualization
- Familiarity with the ongoing and current research in the field of network science
- Understanding of additional (non-network methods) that enable network research, including:

  - Computational statistics (e.g., inferential methods)
  - Data acquisition and handling
  - Measurement and research design

Graduates of the program should also be capable of leading and performing independent, new research projects related to network sciences. Students will need to show competency to communicate network science concepts, processes, and results effectively, both verbally and in writing. It is expected that graduates will be well-prepared to enter many potential career paths including industrial research positions, government consulting positions, postdoctoral researchers, or junior faculty positions in academic institutions.

4 Degree Requirements

Required coursework includes: (1) Three network science foundational courses (Complex Networks and Applications, Network Science Data I, and Dynamical Processes in Complex Networks); (2) data analytic courses (students select either Social Network Analysis or Data Mining Techniques); (3) three to four elective courses (twelve semester hours), defined by their specific track and research goals; and (4) two independent research courses with core faculty of the program. Electives are dependent on a student's area of concentration and subject to approval by their faculty advisor. The expected time to degree is 5 years. Below we give the description of the courses that our PhD students most usually take (required and elective). Additional information on the course syllabi can be found here: https://www.networkscienceinstitute.org/phd.

4.1 Required Core Courses

1. Complex Networks and Applications

Introduces network science and a set of analytical, numerical, and modeling tools used to understand complex networks in nature and technology. Focus is on the organizing principles that govern the emergence of real networks and the theoretical concepts necessary to characterize and model them, with examples coming from biology (metabolic, protein interaction networks), computer science (World Wide Web, Internet), and social systems (e-mail, friendship networks). Covers elements of graph theory, statistical physics, biology, and social science as they pertain to the understanding of complex systems.

2. Network Science Data I

An introductory course on programming for network and data scientists. Students learn the fundamentals of computer programming (e.g., control structures, data structures, algorithms) with particular focus on applications to network and data sciences, such as how to create a network and analyze its basic features using Python.

3. Dynamical Processes in Complex Networks

Immerses students in the modeling of dynamical processes in complex networks (contagion, diffusion, routing, consensus formation, etc.). Provides a rationale for understanding the emergence of tipping points and nonlinear properties that often underpin the most interesting characteristics of sociotechnical systems. The course reviews the recent progress in modeling dynamical processes that integrate the complex features and heterogeneities of real-world systems.

4. Data Analytic Courses

a. Social Networks

Offers an overview of social network analytic methods including topics such as how to characterize topology and visualize and analyze networks. Explores major social network research, covering seminal work from political science, sociology, economics, and physics including small-world literature and the spread of information and disease.

b. Data Mining Techniques

Covers various aspects of data mining, including classification, prediction, ensemble methods, association rules, sequence mining, and cluster analysis. The class project involves hands-on practice of mining useful knowledge from a large data set.

5. Independent Research

Offers advanced students an opportunity to work with an individual instructor on a topic related to current research. Instructor and student negotiate a written agreement as to what topics are covered and what written or laboratory work forms the basis for the grade. Viewed as a lead-in to dissertation research.

4.2 Concentration Elective Courses

1. Network Science Data II

Explores advanced topics of network analytical approaches and practical exercises in real network data. Students learn how to retrieve network data from the real world, analyze network structures and properties, study dynamical processes on top of the networks, and visualize networks. Covers topics such as centrality measures, network sampling and filtering, temporal networks, community detection, network visualization, multiplex networks, and big data network analysis.

2. Statistical Physics of Complex Networks

Covers applications of statistical physics to network science. Focuses on maximum-entropy ensembles of networks and on applicability of network models to real networks. Main topics covered include micro-canonical, canonical, and grand-canonical ensembles of networks, exponential random graphs, latent variable network models, graphons, random geometric graphs and other geometric network models, and statistical inference methods using these models. Covers applications of maximum-entropy geometric network models to efficient navigation in real networks, link prediction, and community structure inference.

3. Special Topics: Bayesian and Network Statistics

An introduction to advanced quantitative methods including maximum likelihood, hierarchical models, sampling, and network modeling. The course begins with a review of probability and then examines maximum likelihood methods for estimating regression models with continuous and categorical dependent variables, followed by examining a variety of procedures for sampling from posterior distributions. These methods are applied to hierarchical modeling and other simple probabilistic models. The course then takes a closer look at the statistical modeling of networks as it has been developed in the social sciences, beginning with the exponential random graph model (ERGM) and finishing with the temporal SIENA model.

4. Introduction to Computational Statistics

Introduces the fundamental techniques of quantitative data analysis, ranging from foundational skills to more advanced topics in statistics, machine learning, and network modeling. Emphasizes real-world data and applications using the R statistical computing language. Prepares students to apply a wide variety of analytic methods to data problems, present their results to non-experts, and progress to more advanced coursework.

5. Network Economics

Covers seminal work in the economics of information and networks. Progresses through concepts of information, its value, measurement, and uncertainty; two-sided (or multi-sided) network effects, organizational information processing, learning, and social networks; how rational actors use information for private advantage; and other micro- and macro-economic effects such as matching markets. Students are expected to produce a paper suitable for publication or inclusion in a thesis.

6. Graph Theory

Covers fundamental concepts to include adjacency and incidence matrices, distance in graphs, matchings and factors, connectivity, network flows, vertex colorings, Eulerian circuits and Hamiltonian cycles, planar graphs, and Ramsey theory.

7. Algorithms

Presents the mathematical techniques used for the design and analysis of computer algorithms. Focuses on algorithmic design paradigms and techniques for analyzing the correctness, time, and space complexity of algorithms. Topics may include asymptotic notation, recurrences, loop invariants, Hoare triples, sorting and searching, advanced data structures, lower bounds, hashing, greedy algorithms, dynamic programming, graph algorithms, and NP-completeness.

4.3   Student Requirements & Evaluations

• Credit Hours

A minimum of 32 credit hours of coursework is required, though the graduate program committee may recommend additional coursework based on student research interests. In principle, course requirements can be waived for students transferring from other programs upon the analysis of the transcript by the program director of the network science PhD program in consultation with the graduate program committee. Up to 9 credit hours can be transferred from regionally accredited US graduate programs, with approval of the program director.

• Minimum Academic Standards

Satisfactory progress in the program is ongoing and formally evaluated at the end of both the first and second years of the program. Students are expected to maintain a cumulative GPA of 3.0 or better in all coursework and to earn at least a 3.0 in the two core foundational and two core data analytic classes. A student who does not maintain the 3.0 GPA, or is not making satisfactory progress on their dissertation research, may be recommended for termination by the graduate program committee.

• Dissertation Advising

Each student has one primary research advisor from the network science doctoral program faculty. As part of the admission evaluation, an initial match is made between new incoming students and faculty members of the Institute. However, during their first and second years in the program, students are expected to meet with their assigned advisor, as well as with other institute faculty members, to determine the best advisor-advisee match. Students must solidify the relationship with their selected research advisor by the end of the spring semester of their second year.

- Qualifying Examination

All students are required to take the qualifying exam in the fall semester of their 3rd year of the program. Students receive 50–80 potential questions one month before the exam, which consists of a set of questions provided by each one of the core faculty of the Institute. Students are asked a subset of these questions by the qualifying examination committee, to which they must be prepared to provide in-depth answers in an oral format. Students have up to 2 hours to show competency across the topics, after which the committee meets to evaluate the student's performance. The committee provides feedback immediately to the students, offering suggestions for growth and direction. Students who do not pass the qualifying exam on their first attempt are expected to retake the exam in the spring term. Students may take the qualifying exam no more than twice. Students who fail to complete the qualifying examination but who have completed all the PhD program's required coursework with a cumulative GPA of 3.0 or better will be awarded a terminal Master of Science in network science degree.

- Comprehensive Examination

PhD students must submit a written dissertation proposal to their dissertation committee. The dissertation committee consists of at least four members: the dissertation advisor (tenured/tenure-track Northeastern University faculty member), one additional network science doctoral program faculty member, one expert in the specific topic of research (who can be from outside the university), and one additional tenured/tenure-track faculty member from the concentration department. The proposal should identify relevant literature, define a research problem, outline a research plan, and describe its potential impact on the field. A presentation of the proposal shall be made in an open forum, and the students must successfully defend it before the dissertation committee. The comprehensive exam must precede the final dissertation defense by at least 1 year. Students may repeat the comprehensive examination once if they are unsuccessful in their first attempt.

- Degree Candidacy

A student is considered a PhD candidate upon completion of required coursework with a minimum cumulative GPA of 3.0, satisfactory completion of the qualification examination, and satisfactory completion of the comprehensive examination.

- Dissertation Defense

A PhD student must complete and defend a dissertation that involves original research in network science. The dissertation defense must adhere to the College of Science policies, as outlined in the Northeastern University Graduate Catalog.

4.4   Sample Course Outline

The curriculum is designed to provide PhD students with a strong foundation in network science. Below we present a sample course outline.

Year 1, Fall Semester
Complex Networks and Applications   (4 credits)
Network Science Data I   (4 credits)
Year 1, Spring Semester
Network Science Data II   (4 credits)
Concentration Elective   (3–4 credits)
Year 2, Fall Semester
Dynamical Processes in Complex Networks   (4 credits)
Concentration Elective   (3–4 credits)
Research   (2 credits)
Year 2, Spring Semester
Social Network Analysis   (4 credits)
or
Data Mining   (4 credits)
Concentration Elective   (3–4 credits)
Research   (2 credits)
Dissertation (PhD Candidacy Achieved)
Year 3 and Until Completion
Dissertation Research

5   PhD Program Progress & Evaluation

The PhD program in network science was launched in the fall of 2014. Now in its fourth year, our program has seven (7) students in year 1, six (6) students in year 2, five (5) students in year 3, and three (3) students in year 4. We have brought together promising young scientists from various backgrounds, who are eager to combine their thoughts and basic principles from the different scientific disciplines they come from. As the first US network science PhD program, other organizations and academic colleagues are looking toward our program for ideas about how to develop and structure a graduate level program in network science. Our team is dedicated to leveraging the existing momentum in the field by evaluating and modifying the program when needed. We rely on the input from and dialogue among the diverse core faculty, as well as the active engagement with the student body. The doctoral students have been exceptionally proactive in this endeavor, forming a student council to formally organize activities and assess all aspects of the program. As part of this ongoing dialogue between the leadership and students, we implemented an anonymous online survey. We received fifteen[1] responses from the first three cohorts of the doctoral program.

5.1 Program Evaluation Survey

During the academic year 2016–2017, we sent all 15 students a 7-question survey. Students were asked about both the content and ordering of the courses; the process of selecting an advisor; and about how prepared they feel for the job market. Twelve students (80%) indicated that they would like additional classes to be offered in the program. They mentioned analysis and method courses that they would like to have had the opportunity to take in the first semester of the program, including Advanced Statistics (e.g., causal inference), Experimental Methodology, Linear Algebra, Calculus and Probability, and a few applied courses, most prominently, network neuroscience. Eight students (53.33%) said that they were happy with the sequence of courses, while three students (20%) were neutral, and four students (26.67%) would like some changes. For example, they mentioned that the network science Data course should be offered before or at the same time as the Complex Networks and Applications course. Eight students (53.33%) suggested that the Network Science Data course should be offered in the first semester or even be divided into two courses, introductory and advanced. Students were very positive when asked about some new activities in the program, including a graduate student speaker series. In terms of professional trajectory, eleven students (73.33%) indicated that they would consider positions in academia, ten students (66.67%) are interested to work in the business and corporate environments, while seven students (46.47%) said they would also consider government or nonprofit positions.

One of the most important insights from the survey arose from the widespread concern about how students who come from different fields are differentially prepared for the courses and research tasks. Students suggested a buddy system for incoming students and a weeklong program (Net Sci boot camp) to ensure that future cohorts will be at equivalent levels in basic coding, complex systems, and network theory. In the months following the survey, the students presented to the faculty a proposal for 5-day introductory session designed for the incoming doctoral student cohort and taught by the doctoral students in their second or third year of the program. The ability to both identify core topic areas necessary across disciplinary

---

[1] In the academic year 2016–2017, the cohorts of the first three years were fifteen (15). Specifically, the incoming students in that year were seven (7). All the fifteen students participated in the survey. However, one student of the year 2016–2017 had, for personal reasons, to restart the PhD program in the following academic year, 2017–2018. This is the reason that, in the end, there are six (6) students in year 2 and seven (7) students in year 1.

backgrounds and then to self-organize in such a way to competently teach those topics is a true signal of achievement for our program. Building common language and context represents the greatest challenge of interdisciplinary endeavors. Our students were able to discover this and design a solution in a matter of months, attesting to the power of interdisciplinary thinking and problem-solving.

5.2   Changes Introduced to the PhD Program

In the time between program launch in fall 2014 and 2017, there has been revision to the content and structure of the PhD program. These revisions have been based on continuous and ongoing evaluation of PhD program, to ensure that the educational, research, and professional objectives of the program have the best opportunity for success. The discussions between the faculty members, the instructors, the leaders and the students, as well as the survey results, led the leadership of the PhD program to introduce changes for the Academic Year 2017–2018. These changes are the following:

- There are two new concentration elective courses in the curriculum. One is the Statistical Physics of Complex Networks. The second elective was derived by splitting the original Network Science Data course into two. The Network Science Data I course will now be an introduction to use computational and algorithmic approaches to the analysis of network data and will be taught, as a core course, in the Fall Semester, in accordance with the theory course "Complex Networks and Applications" which the students take at the same time. The Network Science Data II will be offered in the Spring Semester, starting in 2018, and it will cover, as an elective course, more advanced numerical methods of analysis of networks (e.g., centrality measures, network sampling and filtering, temporal networks, community detection, network visualization, multiplex networks, big data network analysis).
- Due to the difference in the backgrounds of the incoming students, the PhD students are self-organized – with the support of program leadership – to carry out a 5-day boot camp series that took place 1 week before the official start of the semester. Each session was led by the older students on scientific principles and concepts required for the successful completion of the regular classes. Sessions included basic programming languages and computing tools, statistical analysis, linear algebra, whiteboarding, approximations and concepts used in physics (regimes, Taylor expansion, mean field, units, variable transformation, master equation), core algorithmic approaches, and origins of network science within the fields of complex systems and social network analysis.
- Beginning in the spring semester of 2017, we have been holding a weekly network science Literature Review Seminar led by a senior faculty member. Each week the group discusses seminal papers in network methods and theories.

5.3 Future Steps

The program has required significant improvisation, reflecting its newness and the interdisciplinary interests of the students. Critical to successful adaptation has been to listen to and work with doctoral students, and we anticipate ongoing evolution in future years. Building in feedback mechanisms, such as the survey, regular town halls with all of the students, and monthly meetings with the leadership of the students, is likely good practice for all doctoral programs but in this case was essential. We will continue to assess the introduction of the specific changes through additional surveys as well as more formal assessments of students performance through graded activities within the classroom and during the qualifying examination. In addition, as the first graduating cohort reaches the job market, we will be in a more suitable position to assess the program in total. At this phase, what we know is the importance of helping the students to overcome the difficulties that the interdisciplinary nature of the program itself creates. Focused on this target, we are waiting for the outcome of the changes made for the academic year 2017–2018, and we are ready to discuss with the students and consider their needs on a continuous basis.

We are encouraged that in addition to the PhD program offered at Northeastern University, there are new programs on network science being developed in the United States and in Europe. Such programs are the following:

- Network Science Certificate program at the Naval Postgraduate School (2013) [4]
- Undergraduate Minor in Network Science at the US Military Academy (2014) [5]
- PhD program in Network Science at the Central European University (2015) [6]
- Advanced Certificate program in Network Science at the Central European University (2015) [7]
- Network Science Traineeship program at UC Santa Barbara (NSF-IGERT, 2015) [8]
- Network Biology Traineeship at University of Maryland (NSF-NRT, 2017) [9]
- Complex Networks and Systems Traineeship at Indiana University (NSF-NRT, 2018) [10]
- Master of Science in Network Science at Queen Mary University, London, UK (2015) [11]
- Master in Complex Systems, "Physics, Computer Science, and Complex Networks," at École Normale Supérieure (ENL), Lyon, France [12]

It is very promising that the scientific community has started to recognize the specific value of training in network science and specifically of creating young researchers educated as network scientists. As additional programs are established, we are optimistic that collaborations and exchange programs can thrive in the future.

In the spirit of building a student community, we organized a workshop for young scientists who are working on network science-based problems. We invited students from all backgrounds to participate in a one-day meeting to share research, discuss interdisciplinary challenges, explore career paths, and engage in dialogue with experts in the field. As part of this initiative, the students involved with the planning also officially formed the Society of Young Network Scientists (SYNS).

The vision for the society is to build a cohesive community among young network science scholars working across disciplines and within a diversity of research centers. The group currently has 40 students, with active social media profiles, consensus on the organizational structure, and plans to host annual meetings as well as smaller workshops throughout the year.

Acknowledgments We thank the PhD students for their continuous interest in the program and in particular Sarah Shugars for the clarification she provided on the progress of the students' efforts.

References

1. Manyika J., Chui M., Brown B., Bughin J., Dobbs R., Roxburgh C., & Byers A. H. (2011), Big data: The next frontier for innovation, competition, and productivity, McKinsey Global Institute, available at https://www.mckinsey.com/business-functions/digital-mckinsey/ourinsights/big-data-the-next-frontier-for-innovation, accessed March 24, 2018.
2. Chui M., Manyika J., Bughin J., Dobbs R., Roxburgh C., Sarrazin H., Sands G. & Westergren M. (2012), The social economy: Unlocking value and productivity through social technologies, McKinsey Global Institute, available at https://www.mckinsey.com/industries/high-tech/ourinsights/the-social-economy, accessed March 24, 2018.
3. National Research Council, (2005), Network Science, Washington, DC: The National Academies Press, available at https://doi.org/10.17226/11516, accessed March 24, 2018.
4. Naval Postgraduate School, Applied Mathematics Department, Academic Certificate in Network Science, available at https://my.nps.edu/web/math/network-science, accessed March 24, 2018.
5. United States Military Academy WEST POINT, Network Science Center, News, USMA's Newest Academic Minor: Network Science, available at https://www.usma.edu/nsc/SitePages/News.aspx, accessed March 24, 2018.
6. Central European University, Center for Network Science, PhD Program in Network Science at CEU, available at https://cns.ceu.edu/phdprogram-network-science-ceu, accessed March 24, 2018.
7. Central European University, Center for Network Science, Advanced Certificate Program in Network Science, available at https://cns.ceu.edu/node/181, accessed March 24, 2018.
8. University of California, Santa Barbara, Network Science, Integrative Graduate Education & Research Traineeship (IGERT), Education and Training, available at https://networkscience.igert.ucsb.edu/education, accessed March 24, 2018.
9. University of Maryland, COMBINE, Computation and Mathematics for Biological Networks, available at http://www.combine.umd.edu/, accessed March 24, 2018.
10. Indiana University Bloomington, Interdisciplinary Training in Complex Networks and Systems, available at https://cns-nrt.indiana.edu/, accessed March 24, 2018.
11. Queen Mary, University of London, Network Science MSc (1 year Full-time / 2 years Part-time), available at https://www.qmul.ac.uk/postgraduate/taught/coursefinder/courses/143463.html, accessed March 24, 2018.
12. École Normale Supérieure de Lyon, Sciences de la Matière, Master 2 training in Complex Systems: "Physics, Computer Science and Complex Networks", available at http://www.ens-lyon.fr/MasterSDM/en/master-2/m2-complex-systems, accessed March 24, 2018.

Europe's First PhD Program in Network Science

János Kertész and Balázs Vedres

1 The Need for a Network Science PhD Program

During the past decade, network science has become a highly fertile interdisciplinary field, integrating natural and social sciences as well as "formal sciences," like mathematics, statistics, and computer science around the coherent agenda of identifying the mechanisms that govern the dynamics of complex interacting systems with a large number of constituents from cell metabolism and brain structure to the emergence of social movements and international trade dynamics.

In fact, the new discipline of network science has emerged and matured remarkably fast. Conference series have been organized, perhaps the most important being NetSci, an annual international gathering of hundreds of researchers from all over the world (with the latest being NetSci 2017 in Indianapolis [1]). These conferences have regular satellite sections on Network Science in Education. Excellent review articles, monographs, and textbooks have been canonizing disciplinary knowledge, as shown by this list, without a claim for completeness [2–9]. A number of journals have been launched [10–13], and centers have emerged in the USA, Europe, and Asia. In 2012, at the Budapest NetSci conference, the Network Science Society [14] was established, stepping into the role of coordinating events within the discipline and serving as an information hub as well. Network science university courses are by now regular at many major universities. Modern network science is growing with no signs of slowing down; results and their broad applications are accumulating faster than ever. In fact, one of the reasons for the rapid development of network science is the versatile applicability of the results of this new discipline.

We are also witnessing in the USA and in most European countries an ever-increasing job market that is keen to absorb highly qualified network scientists. This evolution goes in parallel with and not independently from the penetration of data

J. Kertész (✉) · B. Vedres
Central European University, Budapest, Hungary

© Springer International Publishing AG, part of Springer Nature 2018 87
C. B. Cramer et al. (eds.), *Network Science In Education*,
https://doi.org/10.1007/978-3-319-77237-0_6

science into many fields; however, data scientists and network scientists use, to some extent, complementary approaches. Both disciplines learn from each other, but data science focuses more on algorithms and pattern identification to make predictions, while network science concentrates more on understanding the laws of the behavior of complex systems. Accordingly, experts with a solid background in network science are welcome in social media and telecommunication services. They introduce new, efficient tools in marketing and finance. A whole new branch of business consulting has emerged, building massively on network science. We can confidently predict that the number of openings will increase in such areas as public policy and government and international and civic activism. The spectacular developments in network science have had a major impact on university programs: the resulting research programs and centers (see, e.g., [15–19]) as well as funding for large-scale projects generate attractive job opportunities in academia.

Network science naturally appears in a number of PhD programs; its results form important parts of bioinformatics, sociology, and economics. The new science of networks has revolutionized complexity science by making clear that unfolding the topological aspects of the interaction between the constituents is necessary for the understanding of such systems. Accordingly, considerable parts of network science occur in the corresponding PhD programs (see, e.g., [20–27]). The high level of maturity of network science as a discipline, together with the increasing need for well-educated experts, made launching a new network science PhD program a natural idea. Several centers around the world started simultaneously thinking about such programs. The first one was launched in 2014 at Northeastern University, and the second one (and the first in Europe) was initiated by the Center of Network Science of Central European University in 2015.[1]

2   Problem-Driven Interdisciplinary Approach at CEU

Central European University (CEU) [28] is an English language postgraduate training institution of higher education with accreditation both by the Middle-States Network of American Universities, the New York State Board of Education, and the Hungarian government.[2] From its inception in the early 1990s, CEU has implemented a systematical policy of recruiting its staff and students in a multicultural

---

[1] Networks occur in many PhD programs from abstract graph theory to engineering. An appealing aspect of the network science PhD program is that it gives a solid and broad basis in this subject together with a large variety of applications.

[2] CEU has been under attack by the Hungarian government during the last months, which tried to hamstring the university by a new legislation [29]. This move has launched a wave of protests in Hungary and worldwide [30]. The Venice Commission is dealing with the Lex CEU [31], and the European Commission has started an infringement proceeding in this matter [32]. CEU is determined to continue its mission as a center of academic freedom, and its programs, including the PhD program in network science, will go on without any disruption [33].

environment. As a consequence, the staff of CEU is decisively international, with a number of American, West, and East European scholars as well as some coming from the Mediterranean region, Scandinavia, and the Near East (Turkey, Israel, Egypt), altogether from more than 40 countries. Students from more than 100 countries have earned master's or PhD degrees at CEU. It is a leading and well-established international academic institution known for the high level of interdisciplinary research. CEU's special strengths are in social sciences and humanities; however, it is increasingly focusing on data aspects and quantitative methods.[3]

The Center for Network Science (CNS) [36] was established in 2008 by Balázs Vedres, with strong support from Yehuda Elkana, the president of CEU and Liviu Matei, the provost. CNS was formed with the mission of building academic excellence in the field of network science and strengthening quantitative interdisciplinary research. An important feature of this mission is to deal with network aspects of social problems central to the core values of CEU, such as social inequalities, discrimination, and exclusion; abuses of political and economic power; the emergence of civic activism; the origins of collective creativity; gender inequalities; and the protection of free speech and open communication. The Center aims to fulfill its mission by actively collaborating with (and to some extent integrating into) other departments at CEU, as well as other research centers in Europe and the world. The Center has external members from Mathematics, Sociology, Political Science, Economics, and Environmental Science departments. Among others, the Center organized NetSci 2011 [37] that brought more than 400 participants to CEU.

CNS started to offer a certificate in network science in 2010. This was an add-on specialization to other PhD programs [43]. The idea for a proper PhD program in network science was first raised in 2012, with strong support from CEU president John Shattuck and provost Katalin Farkas and with the active involvement of Albert-László Barabási. The program started in the fall of 2015 with the arrival of its first cohort and with János Kertész as the first program director.

We at CEU see network science as intrinsically multidisciplinary. It has grown out of the mathematical theory of graphs, and it uses computer science tools as well as concepts and methods of statistical physics. Sociology contributed substantially to its development already at an early stage, but economics, political science, and environmental science have also provided important influences. The excellence of the faculty in these disciplines, together with the expertise at the Center of Network Science at CEU and the multi-departmental character of the program ensures a high quality PhD degree in network science.

To better grasp the particular ways in which we translate multidisciplinary ambitions into action, we outline some of our integrated activities below.

We collaborate closely with colleagues in Mathematics. Network science deals with huge databases and often with networks of the order of 10^6 nodes. One of the focus problems in graph theory is the problem of graph limits, which is closely

---

[3] For general ranking data, see [34] and for individual disciplines [35].

related to large networks. A further mathematical problem is related to the dynamics of and on networks, which involve stochastic processes and differential equations. The Department of Mathematics and its Applications together with the Rényi Institute of Mathematics as well as the members of the CNS have the required experience to provide top quality teaching and guidance in these fields.

Economics is one of our key areas for interdisciplinary integration. The economy is a complex system, and a network approach can be fruitful at many levels, representing diverse points of view. Trade networks, ownership networks, membership and affiliation networks, or networks of market players following similar strategies are just some examples. The 2008 global financial crisis shed light on the importance of the topology of financial networks as cascading failures of the interbank loan system depend heavily on it. Analysis and modeling of data from a network point of view will be crucial in understanding financial interactions and systemic risk and in developing efficient regulation [38]. The availability of some detailed data on individual strategies of investors offers an unprecedented depth in the analysis of individual behavior and their clustering. To this end, the methods of statistical validation combined with advanced network theoretical tools have proved to be extremely efficient and generally applicable [39].

We also pursue several projects together with colleagues from political science. Networks undoubtedly play an important role in political interactions [41]. We think that the formation of opinions and the way the process is related to the social structure and finally to the articulation of political will require a theoretical approach based on multi-level dynamic networks. Network science finds applications in numerous areas: the spreading of information and its relation to manipulative techniques, the analysis of bottom-up and top-down organizations and strategies, discourse analysis and its relation to social networks, voting behavior, party structure and development, etc. The study of these phenomena relies on existing techniques, but also requires new tools. The collaboration between CNS and the Department of Political Science thus has a synergic effect.

Environmental science is another key area of interdisciplinary interest for network science at CEU. One of the main issues here is environmental stability, which is closely related to a focal question in network science, the robustness and vulnerability of complex networks. The stability of food webs, biodiversity, energy security, and sustainable development strategies under transitions are all areas with direct network relations [40]. Collaboration between the staff of the Department of Environmental Science and CNS ensures a high-level education and timely PhD work in this field.

Sociology is an obvious area where network science can contribute to the solution of crucial research problems. On one hand, network science has been significantly influenced by social network analysis. On the other hand, the information and communication revolution has opened unprecedented opportunities to approach classical questions of social sciences like identifying the driving forces behind violence or the factors influencing how ideas, attitudes, and prejudices spread through human populations. A new interdisciplinary field called computational social science [42] emerged with strong emphasis on the network aspects. Researchers of CNS have been at the forefront of this scientific endeavor.

Network science is linked to many further activities at CEU. In the study of international relations, the network aspect seems unavoidable. Cognitive science is related to network science in at least two ways: analysis of the network of neurons is one of the biggest challenges in this field. Furthermore, the analysis of a large amount of temporal data massively uses the recent results of network science. Even in the field of history, network science can be helpful, e.g., in the analysis of narrative networks. The network science program plans to build up links to the programs of the corresponding CEU departments through joint research projects.

This interdisciplinary environment, together with the great Hungarian traditions in graph theory and network science, has made CEU an ideal place for a new PhD program in network science.

3 The Program

3.1 Aims

The purpose of the program is to offer doctoral-level education matching the highest international standards, to train researchers who would take advantage of the opportunities created by large demand for network analysis expertise. Our aim is to train researchers with theoretical, mathematical, and computational skills as well as hands-on experience with large data sets and participation in international research projects.

The program is unique and novel in many ways. It is a program that is offered by the Center for Network Science (CNS), in close collaboration with the departments of Mathematics, Economics, Political Science, and Environmental Sciences. Students with extremely diverse backgrounds are invited to apply. They learn the fundamental ideas of network science and possess deep statistical, big data management and modeling skills, and they are trained in applying network science to real-world problems and are equipped to undertake independent research in a wide variety of network science areas, both in academia and industry.

The PhD program in network science is primarily research-oriented and includes substantive training in data analysis methods. It is interdisciplinary, with substantial course work and research collaboration from other departments. The doctoral theses generated within the framework of the proposed network science PhD program should be typically based on closely supervised intensive empirical research studies that require often (though not exclusively) the application of large data set based on computationally sophisticated methodologies. The special requirements to efficiently realize such a research-oriented doctoral program is reflected in the organizational structure of the PhD degree curriculum.

3.2   The Curriculum

The PhD studies are planned for up to 4 years, with 3 years covered by a regular CEU fellowship and 1 year (in between) spent abroad at one of our partner institutions.[4] Students are considered "probationary PhD candidates" before they pass the comprehensive exam.

First year course work   By the end of the first academic year, probationary PhD candidates have to complete 24 course credits[5] by attending courses offered by the network science PhD program and such courses that are cross-listed with this program.

Research workshop and colloquium   Probationary PhD candidates have to regularly attend the research workshop, where faculty and students present and discuss their work in progress. Students are also expected to participate in the colloquium, a series of invited lectures by network scientists visiting the center.

First year work with a research advisor   Every probationary PhD candidate is assigned a research advisor (or early supervisor) by the doctoral committee of the network science PhD program during the second term (winter term). Students are expected to meet advisors regularly (typically once a week). The task of the research advisor is to help the student to identify their research topic, to draw up a structured plan for data collection and research methodology to be used, and to organize and start the empirical research leading to the preparation of the thesis. The advisor should also be regularly consulted during the preparation of the detailed research proposal.

Comprehensive exam   During the spring term of the first academic year, students should take the comprehensive exam for which an examination committee is appointed by the doctoral committee. The comprehensive exam comprises the topics of the mandatory courses.

Detailed research proposal   In the first year, 6 credits are given for the preparation of the detailed research proposal. The network science PhD program is research oriented, in which the PhD thesis is expected in most cases to be based on the results

---

[4] Partner institutes include the Network Science Institute (Northeastern University), Complexity Science Hub (Vienna), Institute for New Economic Thinking (INEC, Oxford University), Oxford Internet Institute (Oxford University), Department of Computer Science (Aalto University Finland), and Observatory of Complex Systems (University of Palermo). CEU has a formal partnership agreement with INEC, and we are planning such agreements with our most important partners.

[5] A course credit equals a 50-min class per week for 12 weeks.

of data collection, analysis, and modeling. Therefore, by the end of their first year, probationary PhD candidates are required to write a detailed research proposal of 10–15 pages. The research proposal should specify the central question(s) to be investigated and the aim of the research, provide a brief review of previous relevant work and methodologies used to investigate the research topic, include a theoretical rationale for the line of research proposed, and discuss the data to be analyzed. Importantly, the proposal should also state the novelty of the planned work. The detailed research proposal has to represent a realistic and specific plan of the thesis research that should be tailored to be realizable within the framework of the PhD studies.

Courses In the first year, students are expected to do mainly course work, prepare their detailed research proposal, and pass the comprehensive exam. Table 1 gives a list of courses. In the fall term, they have to collect 8 credits from mandatory and a minimum of 4 from elective courses and in the winter term another 8 credits from mandatory and at least 2 credits from elected courses. In the second year, they provide teaching assistance in one of the CNS courses.

Table 1 First year course and related work

Courses	Category	Credits
Fall term		
Fundamental ideas in network science	Mandatory	4
Social networks	Mandatory	4
Research workshop	Mandatory	No credits
Scientific programing in Python	Elective	3
Data and network visualization	Elective	2
Agent-based models	Elective	2
Graph theory (Math Dept.)	Elective	3
Large graphs and groups (Math Dept.)	Elective	3
Other courses with agreement of the doctoral committee	Elective	
Winter term		
Structure and dynamics of complex networks	Mandatory	2
Data mining and big data analytics	Mandatory	2
Statistical methods in network science and data analysis	Mandatory	4
Research workshop	Mandatory	No credits
Economic networks	Elective	2
Stochastic processes in nature and society	Elective	2
Other courses with agreement of the doctoral committee	Elective	
Spring term		
Writing of detailed research proposal	Mandatory	6
Network science (course with changing, adapted topic)	Elective	2
Other courses with agreement of the doctoral committee	Elective	

3.3   Student Recruitment and the First Cohorts

We gave considerable thought to the criteria of admission and the recruiting process for our new PhD program. According to the policy of CEU, we "welcome applications from excellent candidates all over the world," and we provide them with a fellowship sufficient to cover living expenses [45]. We take students with an MSc or MA in a wide variety of fields, including math, physics, computer science, sociology, economics, and finance. The electronic application procedure is run by the admission office [45]. Applicants upload full transcripts of their previous studies and diplomas with an official English translation, at least two letters of recommendation with availabilities of the writers of the letters, a CV, a motivation letter with a conception about planned research, and a proof of English proficiency (usually 100+ scores in TOEFL).

The call is announced on the CEU web page and in various media outlets, such as specialized sites for higher education, bulletin boards relevant for network science, Facebook pages, and via Twitter. In addition, we spread the news through personal channels via direct email.

Thus far we have had three recruiting rounds (2015, 2016, and 2017). There has been a marked increase in the number of people interested in our program; currently the ratio of applicants to admitted students is 10:1. Applicants come with very diverse backgrounds from a large number of countries. The procedure is as follows: an ordered short list is selected by the doctoral committee based on the motivation letter, the recommendations, previous studies, and interviews. Applicants that are selected then for admission are notified.

Presently we have 15 students from 10 countries (China, Germany, Hungary, Iran, Italy, Mexico, Palestine, Romania, Serbia, US). They have master's degrees in physics, mathematics, sociology, finance, economics, biology, psychology, and architecture. The gender balance is six female and nine male students. The result is a colorful and vibrant community, and an inspiring atmosphere, where one of the most important learning mechanisms is the interaction among the students. We place a strong emphasis on the importance of community, and we encourage exchanges among students by providing a desk in two open offices at the center, where students can, and do, interact intensely. Students are also connected in the virtual space (by the platform "Slack"), and they are active in blogging about events, new results, and recent developments in the discipline.

After admitting our students, we guide them through a process of selecting a supervisor and a research theme. Students are assigned a "preliminary supervisor" during the winter term of the first year, who help the students in the preparation of the detailed research plan. Preliminary supervisors are from faculty members of the CNS or from other units of CEU that are associated with CNS; in the latter case, there is a co-supervisor from CNS involved. In an ideal case, the preliminary supervisor will become the thesis supervisor after the student passes the comprehensive exam.

Students enjoy much freedom in choosing their thesis topic. The range is extremely broad. Presently students are working on the spreading of financial innovations, gender equality in creative teams, corruption networks, and network representations of probabilistic learning, to name a few themes.

In most cases data collection and handling is a pivotal part of the studies. Such data are collected, for example, from army recruiting files of the US Civil War, from public procurements, and from the online software project hosting site GitHub.

4   First Experiences and Outlook

By now we have been able to reflect on our first experiences with the new program. While our enthusiasm has not abated, some problems have become clear. First of all, while it is very stimulating to have such a diverse collection of students, their differing backgrounds pose a real challenge to the instructor. On one hand, students with a more modest level of mathematical knowledge have difficulties catching up with the more quantitative courses. On the other hand, mathematically adept students – such as those with physics degrees – are challenged by the task of verbally analyzing complex circumstances or working through 60 pages of sociological theory from 1 week to the next. The lesson is that at the beginning we have to pay more attention to equalizing the levels of knowledge of our students. We offer a presession course in math, which is compulsory for new students without sufficient matching course work on their record. The Center for Academic Writing provides excellent assistance in improving the students' skills in structuring academic publications. The main tool of harmonization is, however, a strong interaction among the students. We have achieved a sense of community, where students assist each other relying on their complementary backgrounds.

While it is too early to draw definitive conclusions about this new program, we strongly believe that we are on a good track with it. The increasing interest by our applicants and the emergence of similar programs worldwide reaffirm our impetus. Besides PhD level programs [44, 48], new initiatives at the master's level were started recently at Queen Mary College, London [46], and at ENS Lyon [47]. We have already established close collaboration with some of these centers, and in the future we plan to make the connections even more intense by possible joint training programs and exchange of students. The discipline of network science will ultimately solidify by the work and collegial solidarity of those who have obtained a PhD degree in it.

Acknowledgments   We would like to acknowledge the fruitful collaboration on this program with our colleagues at CEU, especially with Albert-László Barabási, Rosario Mantegna, and Roberta Sinatra as well as the support of Guido Caldarelli, Zoltán Toroczkai, and Alessandro Vespignani during the accreditation process.

References

1. NetSci (2017) International School and Conference on Network Science, Indianapolis, June 19–23, 2017, http://netsci2017.net
2. Albert, R. and Barabási, A-L. (2002). Statistical Mechanics of Complex Networks. Review of Modern Physics, 74. 47
3. Newman, M. (2003). The Structure and Function of Complex Networks. SIAM Review 45. 167.
4. Dorogovtsev, S. and Mendez, J. (2003) *The Evolution of Networks*, Oxford: Oxford University Press.
5. Caldarelli, G. (2007) *Scale-Free Networks: Complex Webs in Nature and Technology.* Oxford: Oxford University Press.
6. Barrat, A., Barthélemy, M., Vespignani, A. (2008) *Dynamical Processes on Complex Networks.* Cambridge: Cambridge University Press.
7. Newman, M. (2010) *Networks: An Introduction.* Oxford: Oxford University Press.
8. Easley, D., and Kleinberg, J. (2010) *Networks, Crowds, and Markets: Reasoning About a Highly Connected World.* Cambridge: Cambridge University Press.
9. Barabási, A-L. (2016) *Network Science.* Cambridge: Cambridge University Press.
10. Journal of Complex Networks. (2012) http://comnet.oxfordjournals.org/. Accessed 1/19/18.
11. Network Science. (2012) https://www.cambridge.org/core/journals/network-science. Accessed 1/19/18
12. IEEE Transactions on Network Science and Engineering. (2014) http://ieeexplore.ieee.org/xpl/RecentIssue.jsp?punumber=6488902. Accessed 1/19/18.
13. Applied Network Science. (2016) http://appliednetsci.springeropen.com/. Accessed 1/19/18.
14. Network Science Society. (2012) http://www.netscisociety.net/. Accessed 1/19/18.
15. University of Notre Dame, Interdisciplinary Center for Network Science and Applications. http://icensa.com/. Accessed 1/19/18.
16. Yale Institute for Network Science. http://yins.yale.edu/. Accessed 1/19/18.
17. Northeastern University, Network Science Institute. https://www.networkscienceinstitute.org/. Accessed 1/19/18.
18. Indiana University, Network Science Institute. https://iuni.iu.edu/. Accessed 1/19/18.
19. Cambridge Network of Networks. https://www.cnn.group.cam.ac.uk/. Accessed 1/19/18.
20. EPSRC Complexity Science Centres for Doctoral Training, https://www.epsrc.ac.uk/skills/students/centres/pre2013/complexity/. Accessed 1/19/18.
21. OxfordMartinProgramonComplexity. http://www.oxfordmartin.ox.ac.uk/research/programmes/complexity/. Accessed 1/19/18.
22. UCL CoMPLEX. http://www.ucl.ac.uk/complex. Accessed 1/19/18.
23. University of Warwick, Complexity Studies. https://www2.warwick.ac.uk/fac/crossfac/complexity/study/phd/. Accessed 1/19/18.
24. University of Bristol, Complexity Studies. http://www.bristol.ac.uk/study/postgraduate/2018/eng/phd-complexity-sciences/. Accessed 1/19/18.
25. University of Southampton. http://www.complexity.ecs.soton.ac.uk/. Accessed 1/19/18.
26. University of Lisbon. http://complexsystemsstudies.eu/. Accessed 1/19/18.
27. Complexity Science Hub Vienna. http://csh.ac.at/index/. Accessed 1/19/18.
28. Central European University. https://www.ceu.edu. Accessed 1/19/18.
29. See: http://time.com/4737948/viktor-orban-hungary-europe/. Accessed 1/19/18.
30. https://www.ceu.edu/category/istandwithceu. Accessed 1/19/18.
31. https://www.ecoi.net/file_upload/1226_1494319877_hungary-ngolaw.pdf. Accessed 1/19/18.
32. http://europa.eu/rapid/press-release MEX-17-1116 en.htm. Accessed 1/19/18.
33. https://www.ceu.edu/node/17998. Accessed 1/19/18.
34. Times Higher Education Ranking, https://www.ceu.edu/article/2017-06-21/ceu-ranked-among-top-200-universities-europe-times-higher-education. Accessed 1/19/18.

35. Top Universities Ranking, https://www.topuniversities.com/universities/central-european-university. Accessed 1/19/18.
36. Center for Network Science at CEU. https://cns.ceu.edu/. Accessed 1/19/18.
37. NetSci 2011, Budapest, http://archive.ceu.hu/events/2011-06-06/netsci2011-the-international-school-and-conference-on-network-science. Accessed 1/19/18.
38. Acemoglu, D., Ozdaglar, A. and Tahbaz-Salehi, A. (2015) Systemic Risk and Stability in Financial Networks. American Economic Review, 105, 564.
39. Tumminello, M., Micciche, S., Lillo, F., Piilo, J. and Mantegna, R. (2011) Statistically Validated Networks in Bipartite Complex Systems. PLoS ONE, 6. e17994.
40. Ings, T., et al. (2009) Ecological networks beyond food webs. Journal of Animal Ecology 78, 253.
41. Victor, J., Montgomery, A., and Lubel, M. (eds.) (2017) The Oxford Handbook of Political Networks, http://www.oxfordhandbooks.com/view/10.1093/oxfordhb/9780190228217.001.0001/oxfordhb-9780190228217. Accessed 1/19/18.
42. Lazeretal, D. (2009) Science 323.Washington, DC: American Association for the Advancement of Science. 721.
43. Advanced Certificate Program in Network Science at CEU. https://courses.ceu.edu/programs/non-degree-certificate/advanced-certificate-network-science. Accessed 1/19/18.
44. Network Science PhD program at Northeastern University. http://www.networkscienceinstitute.org/phd. Accessed 1/19/18.
45. CEU Admission Office. https://www.ceu.edu/admissions. Accessed 1/19/18.
46. MSc program in Network Science at Queen Mary College, University of London, http://www.maths.qmul.ac.uk/prospective-students/msc-mathematics-networks. Accessed 1/19/18.
47. Masters training in Complex Systems at ENS Lyon: Physics computer science and complex networks, http://www.ens-lyon.fr/MasterSDM/en/master-2/m2-complex-systems. Accessed 1/19/18.
48. Network Science IGERT Program at Santa Barbara. http://networkscience.igert.ucsb.edu/. Accessed 1/19/18.

Part III
Education Network Analysis

Mapping the Curricular Structure and Contents of Network Science Courses

Hiroki Sayama

1 Introduction

Network science has grown at a rapid pace over the last few decades, producing several major international conferences, scientific journals, research communities, and even academic degree programs [1]. As it has matured as an established field of research, there are already a number of courses on this topic developed and offered at various higher education institutions, often at postgraduate levels. Those courses are delivered in several different departments/disciplines with their respective emphases, such as mathematics, computer science, physics, sociology, political science, management science, systems science, biology, and medicine, and in other more interdisciplinary settings as well.

In those recently developed network science courses, instructors adopted different approaches with different focus areas and curricular designs, depending on their backgrounds, knowledge, and objectives. It should be of particular interest to the network science community to investigate what are agreed or disagreed upon among those instructors on the choices of topics and the curricular flows that go through those topics in a sequential instruction. To the best of our knowledge, there is no prior literature on such a systematic analysis of network science course contents.

The study presented in this chapter aims to collect and organize the information about a number of existing network science courses, generate "maps" of their curricular structures, and identify a set of commonly used curricular contents and typical flows of instruction that connect those contents. Information about course contents were extracted from the online syllabi or schedules of the network science courses and were modeled as a directed weighted graph, to which several network

H. Sayama (✉)
Binghamton University, State University of New York, Binghamton, NY, USA

© Springer International Publishing AG, part of Springer Nature 2018   101
C. B. Cramer et al. (eds.), *Network Science In Education*,
https://doi.org/10.1007/978-3-319-77237-0_7

analysis methods were applied to reveal underlying curricular structure. Potential directions of further improvement of network science curriculum design are also discussed based on the results.

2 Data Collection

We gathered information about existing network science courses from online sources, using the following two websites as the main starting points: Complexity Explorer https://www.complexityexplorer.org/ and Awesome Network Analysis https://github.com/briatte/awesome-network-analysis. From these websites we collected the syllabi or course schedules of several dozen English-based courses that included topics related to networks. As our objective was to analyze the curricular structure of "network science" as an interdisciplinary field of research, we excluded the following types of courses from our analysis:

- Purely mathematical graph theory courses
- Statistics courses that only briefly included network analysis
- Courses on narrowly defined applications (e.g., political analysis, genomic analysis)
- Special topics/seminar courses

As a result, we selected the 30 courses shown in Table 1 as the data sources for our study.

Data collection was conducted manually by the author in April–May 2016. Network science-related topics were extracted from each of the data sources and were grouped by instructional modules shown in the syllabi/schedules. All of the extracted topics were converted to lowercase letters without diacritics to facilitate text processing. The topics were also often normalized/edited/reworded/aggregated at the discretion of the author, to make the vocabulary consistent throughout the analysis. The cleaned final data set (including the rewording rules used in this study) is available from figshare [8].

3 Methods of Analysis

The topics and their curricular sequences extracted from the course syllabi/schedules were initially represented as a directed multigraph by the following procedure (also see Fig. 1): connect topics that appear in the same curricular module to each other with bidirectional edges, to form a fully connected cluster of topics; and connect topics covered in the previous module to those covered in the subsequent module with directed edges, to represent curricular flows. These steps were repeated for all curricular modules in all of the courses. After this edge construction process was over, multiple edges that shared the same origin-destination pair were replaced by a

Table 1 URLs of 30 courses from which curricular information was collected for this study. The original URLs used for data collection in April–May 2016 are shown here, some of which may have been updated since then or may no longer be available. Note that some institutions are represented multiple times in this list, while others appear only once. This may have a biasing effect on the results of analysis

http://barabasi.com/book/network-science
http://bingweb.binghamton.edu/~sayama/SSIE641/
http://faculty.nps.edu/rgera/MA4404.html
http://hornacek.coa.edu/dave/Teaching/Networks.11/
http://mae.engr.ucdavis.edu/dsouza/mae298
http://networksatharvard.com/
http://ocw.mit.edu/courses/economics/14-15j-networks-fall-2009/
http://ocw.mit.edu/courses/media-arts-and-sciences/ mas-961-networks-complexity-and-its-applications-spring-2011/
http://perso.ens-lyon.fr/marton.karsai/Marton_Karsai/complexnet.html
https://cns.ceu.edu/node/31544
https://cns.ceu.edu/node/31545
https://cns.ceu.edu/node/38501
https://courses.cit.cornell.edu/info2040_2015fa/
https://iu.instructure.com/courses/1491418/assignments/syllabus
https://sites.google.com/a/yale.edu/462-562-graphs-and-networks/
https://www0.maths.ox.ac.uk/courses/course/28833/synopsis
https://www.coursera.org/course/sna
https://www.sg.ethz.ch/media/medialibrary/2014/11/syllabus-cn15.pdf
http://tuvalu.santafe.edu/~aaronc/courses/5352/
http://web.stanford.edu/class/cs224w/handouts.html
http://web.stanford.edu/~jugander/mse334/
http://www2.warwick.ac.uk/fac/cross_fac/complexity/study/msc_and_phd/co901/
http://www.ait-budapest.com/structure-and-dynamics-of-complex-networks
http://www.cabdyn.ox.ac.uk/Network%20Courses/SNA_Handbook%202013-14.pdf
http://www.cc.gatech.edu/~dovrolis/Courses/NetSci/
http://www.columbia.edu/itc/sociology/watts/w3233/
http://www.cse.unr.edu/~mgunes/cs765/
http://www-personal.umich.edu/~mejn/courses/2015/cscs535/index.html
http://www.stanford.edu/~jacksonm/291syllabus.pdf
http://www.uvm.edu/~pdodds/teaching/courses/2016-01UVM-303/

single directed weighted edge with the multiplicity of the original edges as the weight. The result was obtained as a single large directed weighted graph, which we call the *topic network* hereafter. This topic network was analyzed using several different methods.

First, the distribution of instructional attention/emphasis in the current network science courses was characterized by measuring the absolute frequencies of appearance of topics in the original data set. We did not use degree or other centrality

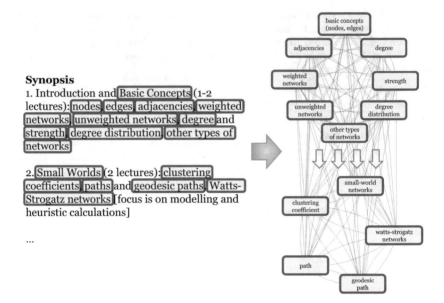

Synopsis
1. Introduction and Basic Concepts (1-2 lectures): nodes, edges, adjacencies, weighted networks, unweighted networks, degree and strength, degree distribution, other types of networks

2. Small Worlds (2 lectures): clustering coefficients, paths and geodesic paths, Watts-Strogatz networks [focus is on modelling and heuristic calculations]

...

Fig. 1 Schematic illustration showing how the edges in the topic network were created from course syllabi/schedules. Left: an excerpt from a sample network science course syllabus (from Mason Porter's course https://www0.maths.ox.ac.uk/courses/course/28833/synopsis; also see [7]), in which extracted topics are highlighted. Right: a subgraph of the topic network created from the excerpt on the left. Topics that appear in the same curricular module were connected to each other with bidirectional edges. Directed edges were also created from topics covered in the previous module (top) to those covered in the subsequent module (bottom) to represent curricular flows. The extracted topics were often normalized/edited/reworded/aggregated at the discretion of the author, to make the vocabulary consistent throughout the analysis

measures in the topic network for this purpose, because, according to the procedure of network construction used in this study (Fig. 1), each topic's in- and out-degrees are greatly influenced by the numbers of other topics in previous and next curricular modules, respectively.

Next, all of the edges whose weight was two or below were removed from the topic network, and only the largest strongly connected component was kept for the rest of the analysis, in order to improve the robustness of the findings by focusing on the essential main body of the topic network. Communities of topics were detected by applying the modularity maximization method [2, 4] to the topic network. Finally, the edge weights were inverted from the original ones so they would represent distance (not strength) of connections, and then the minimum spanning tree (i.e., a tree that reaches all of the nodes with the minimal sum of edge weights) [5] of this weight-inverted topic network was computed to reveal typical flows of instruction going through various network science topics. For all of these analyses and visualizations, we used Wolfram Research Mathematica 11.1.1.

4 Results

Figure 2 shows the top 20 topics that appeared most frequently in the collected course syllabi/schedules. The topic "small-world networks" appeared most frequently in our analysis, probably because this topic was covered widely in various disciplines, including math/physics/computer science, social/economic/political sciences, psychology/neuroscience, and some others. "Random networks," "centrality," and other well-known topics are also represented in this list. A larger set of topics is visualized as a word cloud in Fig. 3.

Figure 4 shows a visualization of the filtered topic network after edge weight thresholding and extraction of the largest strongly connected component. High-resolution versions of this and other visualizations are available from figshare [8]. While the original topic network included 505 topics, the filtered one included 121. The latter was more focused on essential, frequently covered topics than the original, and thus we used the filtered one for the rest of the analysis.Figure 5 shows the communities of topics detected by applying the modularity maximization method to the filtered topic network. Seven topic clusters were detected. Although characterizing each cluster with an appropriate label was a challenging task, we reviewed the

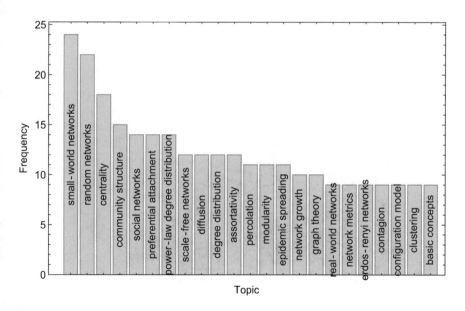

Fig. 2 Frequencies of the top 20 topics that appeared most frequently in the 30 course syllabi/schedules (ties were included so a total of 23 topics appear in this chart). Note that all of the extracted topics were converted to lowercase letters without diacritics to facilitate text processing (this applies to the other figures as well)

Fig. 3 Visualization of topic frequencies in the 30 course syllabi/schedules as a word cloud. Font sizes are set proportional to the square roots of topic frequencies

Fig. 4 Visualization of the topic network after edge weight thresholding. Only the largest strongly connected component is shown. Edge weights are ignored to simplify the visualization

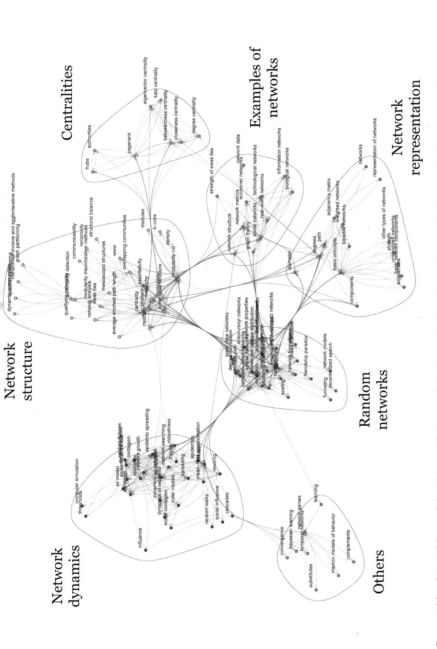

Fig. 5 Communities detected by applying the modularity maximization method to the topic network. Seven topic clusters were detected: (1) *examples of networks*, (2) *network representation*, (3) *random networks*, (4) *network structure*, (5) *centralities*, (6) *network dynamics*, and (7) *others*.

content of each cluster and came up with the following characterization of the seven clusters:

1. *Examples of networks (middle-right).* This cluster includes concrete examples of networks, such as social networks, economic networks, biological networks, technological networks, and information networks.
2. *Network representation (bottom-right).* This cluster includes fundamental concepts and terminologies about representation of networks, such as basic network components, adjacencies, path, degree, strength, etc.
3. *Random networks (bottom-center).* This cluster is the most dense and the most difficult to label. It includes a wide variety of topics, and many of them had strong connections to other communities. However, it uniquely includes several major random network models (e.g., Erdős–Rényi networks, small-world networks, Barabási–Albert networks, preferential attachment, etc.). Therefore we tentatively call this cluster "random networks." It is clearly the core part of this topic community map.
4. *Network structure (top-center).* This cluster includes concepts about network structure and tools to analyze it, such as clustering, path length, modularity, community detection, k-core, etc.
5. *Centralities (top-right).* This relatively small cluster has a clear focus on centrality measures.
6. *Network dynamics (top-left).* This cluster includes various dynamical concepts that are typically discussed in dynamical systems, stochastic/probabilistic systems, and statistical physics, such as spreading/contagion, influence, and dynamics on/of networks.
7. *Others (bottom-left).* This small cluster includes miscellaneous topics that do not appear to have a common theme (e.g., learning, network games, temporal networks).

The cluster of random networks occupies a central position in this map, to which most other clusters are attached with varying degrees of connection strength. The connections are particularly strong between random networks and network structure, as well as between random networks and network dynamics, indicating their strong linkages in the core curricula of network science courses.

We compared the topic clusters identified above with the essential concepts generated by students and educators through the Network Literacy initiative [6, 9] (Table 2). They matched reasonably regarding examples of networks, network representation, network structure/centralities, and network dynamics. In the meantime, the cluster of random networks does not have a counterpart in the essential concepts list, probably because the topics covered in this cluster are somewhat at advanced levels and may not be suitable for secondary education or general public. On the other hand, the essential concepts about visualization and computer technology (4 and 5 in the second column of Table 2) were not well represented in the topic communities seen in Fig. 5. This finding coincides with the fact that those two essential concepts were suggested and emphasized by NetSci High [3] students, not by network science researchers, when the Network Literacy booklet was developed [9].

Table 2 Comparison between the topic clusters revealed in Fig. 5 and the essential concepts developed in the Network Literacy initiative [6, 9]

Topic cluster detected	Essential concept given in Network Literacy	Matched?
1. Examples of networks	1. Networks are everywhere	Yes
2. Network representation	2. Networks describe how things connect and interact	Yes
3. Random networks	(*missing*)	No
4. Network structure	3. Networks can help reveal patterns	Yes
5. Centralities	(*covered in 3*)	Yes
(*missing*)	4. Visualizations can help provide an understanding of networks	No
(*missing*)	5. Today's computer technology allows you to study real-world networks	No
(*covered in 1?*)	6. Networks help you to compare a wide variety of systems	Yes?
6. Network dynamics	7. The structure of a network can influence its state and vice versa	Yes
7. Others	(*covered in 7?*)	Yes?

This may indicate that the current curricular structure of network science courses is likely not spending sufficient time or resource to cover computational tools and visualization methods, even though they could be essential for students' learning of networks. A potential factor contributing to this gap may be that many of the courses analyzed here are at advanced graduate levels, where computational methods and visualization tools may not be part of the core curricular content.

Finally, Fig. 6 presents the minimum spanning tree of the topic network with inverted edge weights. This map shows the curricular structure of network science courses in greater detail with sequential relationships, revealing a possible "backbone" of curricular flows among various network science concepts. The root of the tree is located near the right side of the map, starting with social networks. From there, several curricular flows can be identified on this map. Details are explained below with enlarged portions of the map, which turn out to bear a good correspondence with the topic clusters detected in Fig. 5.

Figure 7 shows the right portion of Fig. 6, in which the root of the spanning tree, social networks, is located in the middle. Two branches are shown in this figure, in addition to another path going from the root leftward. The first branch (lower one) includes topics such as technological networks, information networks, biological networks, and real-world networks, which clearly correspond to the topic cluster of *examples of networks*. The other branch (upper one) includes network data, community detection, partitioning, and other related topics, which could be summarized as *network structure*, together with a few other topics that show up at the tip of the first branch. This area of the tree appears to be an introductory part of the curricular structure.

Figure 8 shows the central portion of Fig. 6, which is the busiest area in the spanning tree where a number of new concepts and models are introduced. The curricular flow that originated in the root comes from the right, and first goes

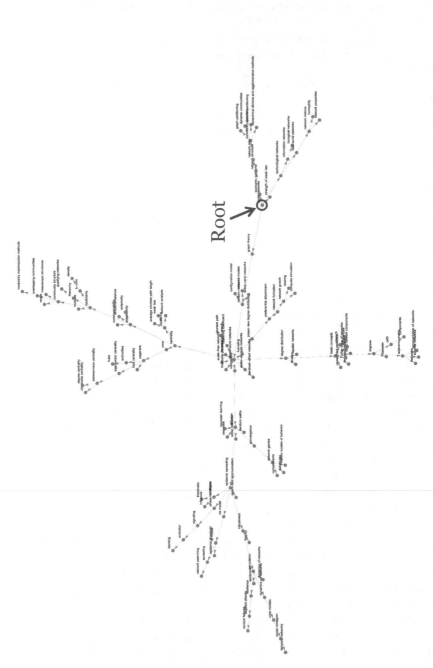

Fig. 6 Minimum spanning tree of the topic network with inverted edge weights. The root of the tree is indicated by a red circle. This spanning tree presents a sample "backbone" of curricular flows among various network science concepts. High-resolution version is available from figshare [8]

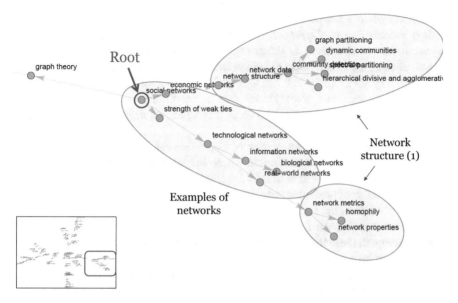

Fig. 7 Enlarged right portion of the spanning tree shown in Fig. 6. Two branches, covering *examples of networks* and *network structure*, extend from the root of the spanning tree

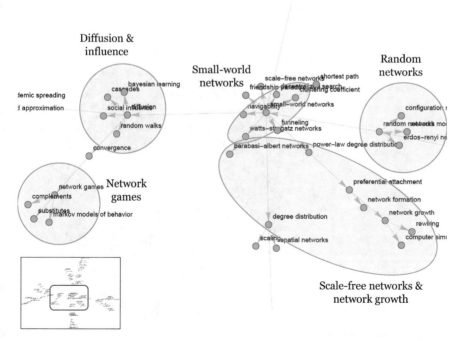

Fig. 8 Enlarged central portion of the spanning tree shown in Fig. 6. The curricular flow originating at the root (not shown in this figure) comes from the right, goes through *random networks*, then reaches *small-world networks*. From there several outgoing branches emanate, including *scale-free networks & network growth* and *diffusion & influence*; the latter is followed by *network games*

through *random networks*, where purely random network models such as Erdős–Rényi models and configuration models are introduced. Then it reaches *small-world networks* that serve as the crux of the whole spanning tree. The observed importance of the small-world networks in the curricular flow agrees with its highest frequency seen in Fig. 2. From there, several different flows branch off toward various subtopics, most notably *scale-free networks & network growth* that goes down to the right. Other topics shown in this figure are *diffusion & influence* and *network games* to the left.

Figure 9 shows the bottom portion of Fig. 6 that can be summarized as a single branch about *network representation*, where fundamental concepts and terminologies about representation of networks are covered, such as degrees, strengths, adjacencies, unweighted/weighted networks, path, diameter, and bipartite networks.

Figure 10 shows the top portion of Fig. 6, which includes a branch for *centralities* and another branch for *network structure*. Together with the bottom branch shown in Fig. 9, these three branches cover various topics about theories and methods of structural analysis of networks.

Finally, Fig. 11 shows the left portion of Fig. 6, which can be considered a large branch of *network dynamics*. Extending from *diffusion & influence* in Fig. 8, this branch covers topics such as epidemic spreading, phase transitions, robustness, percolation, and dynamics on/of networks. It is apparent that this area is predominantly

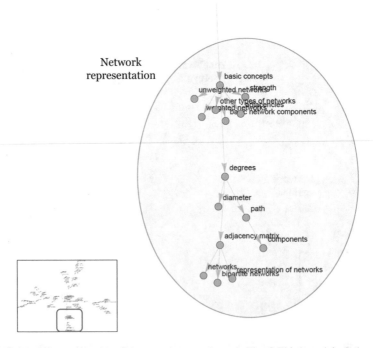

Fig. 9 Enlarged bottom portion of the spanning tree shown in Fig. 6. This branch includes various topics about *network representation*

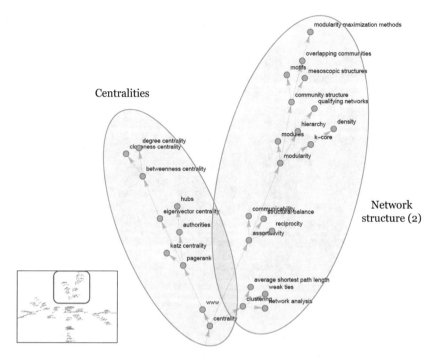

Fig. 10 Enlarged top portion of the spanning tree shown in Fig. 6. This portion first creates a major branch of *centralities* and then creates another on *network structure* that covers topics such as assortativity, modularity, and community structure

oriented to dynamical systems, stochastic/probabilistic systems, and statistical physics, where many advanced concepts, theoretical models, and analytical methods are discussed.

Overall, the examination of the spanning tree illustrated the following steps as a potential curricular flow of network science courses:

1. Start with examples of networks (e.g., social networks), with some basics of network structure.
2. Introduce random networks and small-world networks.
3. From there, take any of the following subtopic paths depending on the objective and need of the course:

   (a) Scale-free networks and network growth
   (b) Network representation
   (c) Centralities
   (d) Other topics on network structure
   (e) Network dynamics

Needless to say, this presents nothing more than just one example of a number of possible instruction designs in teaching network science. Many of the courses

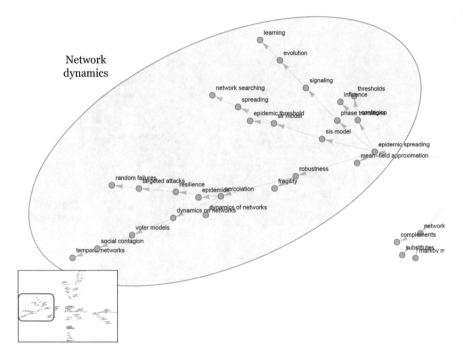

Fig. 11 Enlarged left portion of the spanning tree shown in Fig. 6. This portion, coming from *diffusion & influence* in Fig. 8, includes a wide variety of topics about *network dynamics*, such as epidemic spreading, phase transitions, robustness, percolation, and dynamics on/of networks

included in the dataset of this study adopted a curricular flow substantially different from the one shown above (e.g., see [7]). The curricular flow of a specific course should be carefully custom-designed according to the objective and scope of the course, the academic level and background of students, time/resource constraints, and many other variables.

5 Conclusions

In this study, we constructed and analyzed networked maps of topics covered in 30 existing network science courses. The communities identified in the topic network revealed seven major topic clusters: examples of networks, network representation, random networks, network structure, centralities, network dynamics, and others. These detected clusters showed a reasonable level of agreement with the essential concepts identified in the Network Literacy initiative, although the importance of visualization and computer technology was not well represented in the current network science courses. This presents potential room for instructional redesign; increasing the time and resource allocated for visualization and computer technology may improve students' learning of networks.

We also computed the minimum spanning tree of the topic network to elucidate instructional flows of curricular contents. This analysis revealed a more fine-grained, directed structure of the topic network, in which a typical flow of instruction starts with examples of networks, moves onto random networks and small-world networks, and then branches off to various subtopics from there. This directed topic map will be useful for instructors to navigate through various network science topics and design their own curricula when teaching network science. We hope that the results presented in this chapter offers the first step to illustrate the current state of consensus formation (including variations and disagreements) in the network science community, on what should be taught about networks and how. They may also be informative for K–12 education and informal education as well, when educators and students explore network science topics to choose relevant teaching/learning materials for their needs.

It should be noted that our results depend on the specific choices we made about data sources and data cleaning/analysis methods, which were not fully validated in an objective manner. Conducting a similar analysis using different sources and methods may thus produce significantly different maps of curricular contents. Moreover, as the educational effort of network science has been growing rapidly [10], new courses are continuously created and offered with new topics, instructional designs, and methodologies each year. We suggest that the network science community should continue modeling and analyzing the curricular structure of network science courses in the coming decades, to develop, assess, and adjust effective teaching strategies and methods for this quickly evolving field of interdisciplinary research.

Acknowledgments The author thanks Mason Porter for reviewing the earlier version of this chapter and providing many valuable comments and suggestions, which have significantly helped improve the content.

References

1. Barabási, A.L.: Network Science. Cambridge University Press (2016). Available online at http://networksciencebook.com/
2. Blondel, V.D., Guillaume, J.L., Lambiotte, R., Lefebvre, E.: Fast unfolding of communities in large networks. Journal of Statistical Mechanics: Theory and Experiment 2008(10), P10,008 (2008)
3. Cramer, C., Sheetz, L., Sayama, H., Trunfio, P., Stanley, H.E., Uzzo, S.: NetSci High: Bringing network science research to high schools. In: Complex Networks VI, pp. 209–218. Springer (2015)
4. Fortunato, S., Hric, D.: Community detection in networks: A user guide. Physics Reports 659, 1–44 (2016)
5. Graham, R.L., Hell, P.: On the history of the minimum spanning tree problem. Annals of the History of Computing 7(1), 43–57 (1985)
6. NetSciEd: Network Literacy: Essential Concepts and Core Ideas. http://tinyurl.com/networkliteracy (2015). [Online; accessed July-30-2017]

7. Porter, M.A.: An undergraduate mathematics course on networks. In: C. Cramer, M.A. Porter, H. Sayama, L. Sheetz, S.M. Uzzo (eds.) Network Science In Education – Transformational Approaches in Teaching and Learning. Springer Cham (2018); Available online at: https://www.math.ucla.edu/.mason/papers/net_course04-clean.pdf

8. Sayama, H.: "Mapping the curricular structure and contents of network science courses": Dataset and high resolution figures. https://doi.org/10.6084/m9.figshare.5500843.v1 (2017). figshare [Online; accessed October-14-2017]

9. Sayama, H., Cramer, C., Porter, M.A., Sheetz, L.,Uzzo, S.: What are essential concepts about networks? Journal of Complex Networks 4(3), 457–474 (2016)

10. Sayama, H., Cramer, C., Sheetz, L., Uzzo, S.: NetSciEd: Network science and education for the interconnected world. Complicity: An International of Complexity and Education 14(2), 104–115 (2017). Available online at https://journals.library.ualberta.ca/complicity/index.php/complicity/article/view/29339

Pay, Position, and Partnership: Exploring Capital Resources Among a School District Leadership Team

Alan J. Daly, Yi-Hwa Liou, and Peter Bjorklund Jr

1 Introduction

Capital assets are consequential to individuals' income attainment, particularly around the interplay between human and social capital [1]. Among these capital assets, prior studies indicate that social capital plays a greater role than human capital in determining the level of income [2]. The effect of human capital (e.g., seniority, years of formal education, work-related experience, etc.) on the salary is minor when individuals have access to social capital [ibid]. This view of income attainment underscores the crucial role of social capital, which posits that people's personal network of connections, interpreted as her/his social capital resources [3, 4], is an important means to their chances for improved life outcomes such as labor market opportunities and socioeconomic status. Simply stated, people have access to many more choices and opportunities when they have better relational resources at their disposal [2]. Therefore, the structure of individual networks deserves further investigation in order to better understand the interplay between network position and income as well as the implications for resource allocation in a social system with limited and competitive resources such as a school district.

Studies in economics and social psychology identify several factors that are related to income level as well as network position. However, the link between income, social network position, and factors that may contribute to individual outcomes is limited. In this chapter, we aim to address this gap by exploring the relationships between social network position, income, organizational commitment, and gender. The latter two factors are significant to social embeddedness of individual actors, meaning interpersonal interactions may be related to a person's

A. J. Daly · P. Bjorklund Jr
University of California, San Diego, San Diego, CA, USA

Y.-H. Liou (✉)
National Taipei University of Education, Taipei, Taiwan

© Springer International Publishing AG, part of Springer Nature 2018
C. B. Cramer et al. (eds.), *Network Science In Education*,
https://doi.org/10.1007/978-3-319-77237-0_8

sense of commitment, belonging, and demographics [5]. Specifically, we will examine differences in network position as related to the level of income, organizational commitment, and gender. Our data comes from a group of school principals working in a midsized public school district in southern California that serves a large student population within a socioeconomically, racially, and linguistically diverse system. This chapter will add to the existing knowledge base regarding social networks in education. In the next section of the paper, we will describe our conceptual framework with a brief discussion of the concept of intellectual capital and its relation to commitment, income, and gender. We will then explain the data and methods of the study, followed by the results section, and then we will close with a discussion of our findings.

2   Conceptual Framework

2.1   *Intellectual Capital*

The concept of intellectual capital derives from theories of social networks and social capital [6]. It has been used in fields such as business entrepreneurship, social media, politics, and communication. Across the disciplines and research, it is commonly understood that individual actors develop, maintain, or accumulate social capital by accessing relational resources that are embedded in their personal network [7–9]. Individual actors interact with others in the network for various purposes (e.g., friendship, social support, work-related information, etc.), and they form and/or dispose of their ties with others. This process of interaction places actors in certain positions within networks, which is consequential to the development of their intellectual capital. In this regard, two major concepts of intellectual capital deserve greater attention: relational ties and resulting network position.

Relational ties between actors can be regarded as multiple channels and opportunities in which resources (e.g., support, advice, knowledge, and information) travel across the network [7]. Ties that are mutual are considered reciprocal. Reciprocal ties allow for the efficient transmission of more tacit and complex knowledge and information [10]. These mutual ties help create trust and strengthen social bonds [11, 12], but the flow of resources between ties is also susceptible to stagnation, as information may circulate back and forth within the closed circle of mutually connected actors, create redundancy, and limit the creation of new knowledge [3].

Network position refers to a specific set of incoming and/or outgoing relational ties an actor has within a social network [13]. Actors who occupy better network positions are better able to access and mobilize relational resources [7]. Several types of network positions have been studied in earlier work such as brokers and central actors, each with its purpose of connection and communication. This study attempts to explore the degree of actor connectedness and its relationship to income. We focus on "closeness" of an actors' network position as it measures both the quantity and quality (distance) of ties an actor has. Actors with greater closeness are those individuals who possess the capacity to efficiently and quickly connect with others due to their pattern of relationships and thus occupy advantageous network positions. Greater closeness is associated with power and influence, [14–16], as well as goal attainment and reputational effectiveness of organizational members [17, 18].

2.2 Salary and Social Network Position

Individual salary is associated with investment in human capital as one develops and accumulates knowledge, skills, and ability he/she may qualify for corresponding higher paying careers in organizations [19, 20]. Investment in human capital is usually measured by education degree, seniority, and work experience [2], and often supports career advancement when coupled with social capital [1], and, as a result, the return on investment in human capital may be associated with increased income from work [19–21].

According to Bourdieu [22] and Coleman [4], social capital provides opportunity to garner returns from the application of human capital. For instance, parents with a greater volume of social capital assets can provide better educational opportunities for their children [4], which may increase the child's opportunity to obtain a high-pay position in the future (social mobility). Another example has to do with one's social contacts in job searches. Networks of important contacts often require a certain level of educational background, work status, and experience. Individuals who are able to develop social ties/contacts with important actors who occupy good positions in their field have an advantage when searching for a new job. People with ties are more likely to share these "important" social contacts with one another [23]. The expansion and use of social ties or contacts may increase in both intellectual and financial capital.

Beyond finding new jobs, higher levels of social capital are often related to recruitment for managerial positions [1]. This is because actors with strong ties share higher levels of trust as well as similar norms and beliefs, based on which employers would consider these actors to be loyal and reflect or reinforce the social norms of their close-knit group [1]. Hiring from strong ties may facilitate the exchange of resources and also mitigate potential risk in a partnership, in that the employers and job applicants share mutual trust [24]. Applicants with strong ties with employers possess bargaining power for wages and benefits from this vantage point [25]. As such, building cohesive networks of strong ties is likely to be the means to obtain and sustain a high salary in compensation for the trustworthiness of applicants [1]. This research underlines the important role of intellectual capital. Given that intellectual capital may be determinant of individual actors' job title and resulting salary in a host of organizations and fields, it is reasonable to assume this is the case for the leadership position in education. Therefore, we hypothesize that *school principals with higher salary are more likely to possess central network positions and mutual ties* (hypothesis 1).

2.3   Organizational Commitment and Network Position

Organizational commitment has long been studied in social psychology and business [26–29]. Despite various approaches to understanding the concept, organizational commitment can be generally defined as a psychological state that is associated with the degree of individuals' attachment/bond to her/his organization [28, 30, 31]. Committed employees have a stronger desire to stay in the organization [ibid], whereas less committed employees are likely to display withdrawal behaviors (e.g., absenteeism) and leave the organization [32]. A strong commitment from employees helps increase one's sense of belonging and self-fulfillment [29], job performance and productivity [33], contribution to organizational innovation [34], and reduce organizational turnover [35]. This research examines the commitment from a social psychological model using constructs like sense of work-related competence, group cohesiveness, and work autonomy. Moreover, some scholars correlate commitment with job involvement, job satisfaction, or job stress. Among the extensive literature, the dominant approach to studying organizational commitment treats the concept as a social psychological construct. However, this approach has not sufficiently addressed the "social" aspect among organizational members, given that individuals are embedded within social relationships in her/his organization [36, 37]. This chapter attempts to address this missing link between organizational commitment and social networks.

Network scholars posit that social ties act as relational sources of organizational commitment that bond employees to their organization [38]. Individuals' sense of commitment is generally formed through the desire to accomplish tasks and achieve goals by interacting and collaborating with other organizational members [39–42]. In this regard, organizational commitment is influenced and can be shaped by in the exchange/interaction nexus among organizational members. A number of earlier studies examined the relationships between social embeddedness and organizational commitment. For instance, Roberts and O'Reilly [43] found that employees who were disconnected with other members in an organization's communication network were less satisfied with their work environment than those that are well-connected. Similarly, Brass [44] found that employees possessing central network positions reported less satisfaction with their job. Krackhardt and Porter [45] found that employees who left the organization tended to possess similar network positions (e.g., shared social group) in an organization's communication network, suggesting an association between organizational commitment with network position [46]. Other researchers point to the fact that individuals dissimilar to the rest of the team members tend to exit the team [47]. McPherson, Popielarz, and Drobnic [48] suggested that actors with strong ties (mutual confiding) are less likely to leave a group. Given the interplay between individuals' sense of organizational commitment and their social network positions, we hypothesize that *school principals with higher levels of organizational commitment are more likely to possess central network positions and mutual ties* (hypothesis 2).

2.4 Gender and Network Position

Literature across various disciplines indicates that gender differences in resources and rewards is due in large part to social network position, access to power and resources, and social embeddedness (e.g., Hultin and Szulkin [49]). This literature has shown that females and males have differential access to networks of social contacts and power resources, which in turn affect the mobilization and distribution of power, position, and payment in organizations [12, 50, 51]. This gender differentiation is also based on the assumption that people with same gender are more likely to form/share social ties than with people of a different gender [52–54] and that same-gender work ties tend to be more frequent than different-gender ties [11, 12, 55].

Network literature further demonstrates gender differentiation on social network position. According to the principal of homophily [56], same-gender actors tend to form ties and cohesive groups in that such demographical similarity reduces cognitive dissonance, facilitates communication, and in turn helps establish mutual trust and reciprocity [11, 12]. Studies have shown that women are more likely than men to engage in social relationships [57] and that men are more likely than women to establish professional social contacts at and for work [58, 59]. Furthermore, women are found to be less central than men in organizational networks of power and authority in which important decisions and policies are made [51]. In this regard, women may have less professional support at work in an organization in which men represent authoritative figures and dominate the power in decision-making. Given gender similarity in the occurrence of interpersonal relationships and that men are more likely than women to establish professional networks of work-related contacts [51, 60], we hypothesize that *male principals are more likely than female principals to occupy central network position and possess mutual ties* (hypothesis 3).

3 Data and Methods

3.1 Sample and Context

This study is part of the larger district-wide research project that attempts to understand the alignment between reform efforts and the goals of the district. The present study focuses on the role of school principals as they have direct and indirect influence on school outcomes [61]. The sample of this study includes 29 school principals in one urban fringe school district that serves disadvantaged student populations in socioeconomic background, race/ethnicity, and English language learning status in California. We selected this district as it provides a state-level representative case that reflects the demographic composition of general school districts in California. The district has undertaken a series of improvement efforts with an aim to cultivate a district-wide collaborative community and to achieve innovation. One common leadership practice is the ongoing monthly leadership team meetings in which all district and site leaders gather together to discuss goals, strategic plans, and progress for improvement. It is expected that the leadership team serves as the board of directors of the district organization and that the school principals would work as a team on a regular basis in communicating, collaborating, and exchanging ideas/practices that are tailored to meet the goals of the district.

Of all the 29 principals, 62.1% were female and 72.4% were White, and more than half of the principals had a master's degree. Regarding experience, about two-thirds of the principals were classroom teachers for more than 17 years and have been administrators for more than 6 years. In terms of annual base income, more than two-thirds of the principals receive more than approximately one hundred thousand dollars per year. The sample demographics are presented in Table 1.

3.2 Instruments

We collected data in 2013 from the 29 school principals that included their perceptions of organizational commitment, a set of social network data with regard to the degree of their connectedness with their principal colleagues, and demographic

Table 1 Sample demographics of school principals

	Freq.	%
Gender		
Female	18	62.1
Male	11	37.9
Race/ethnicity		
Hispanic/Latino	4	13.8
White	21	72.4
Multicultural	4	13.8
Degree		
Masters	6	20.7
Masters+30	16	55.2
Doctorate	3	10.3
Years of being an educator		
≤16 years	9	31.0
17–20 years	10	34.5
≥21 years	10	34.5
Years in administration		
≤5 years	9	31.0
6–9 years	10	34.5
≥10 years	10	34.5
Annual base income		
≤91,000	8	27.6
92,000–110,000	11	37.9
≥111,000	10	34.5

information such as gender and years of experience. We also collected data that showed each principals' annual income from publicly available databases on Transparent California. We use individual leader's total pay plus benefits as their annual income base.

Social networks We collected data about the relationships among site leaders using a social network survey. Based on earlier work [62], we developed and validated network questions that would assess the principal networks with regard to leadership advice and work recognition. We asked participants to assess the frequency of interaction with other principals "to whom do you go to for advice on how to strengthen your leadership practice?" (leadership advice (LA) network) and "by whom do you get recognized for your efforts?" (work recognition (WR) network) on a four-point frequency scale (1 = few times a year to 4 = daily). Respondents could indicate with whom they seek leadership advice and by whom they receive work recognition by selecting any of the names of their fellow principals from a roster of names of other principals in the district and assessing corresponding frequency of interaction with each nominee. A bounded approach is appropriate as the current study is focused on the degree of connectedness within a finite network of principals [63]. Further, as we attempt to explore principals' network position along with factors that may be attributing to the position, we focus on two-plex network relation and its resulting network position. The two-plex relation, which we refer to as LAWR relation in this chapter, indicates a joint phenomenon in which a leadership advice coexists with a work recognition in the same direction. For instance, if principal A receives a LAWR tie from principal B, this indicates that A is regarded by B as a provider of both leadership advice *and* work recognition. We argue that such relation may represent a certain degree of important relational ties and will allow us to explore relationships between principals' network position, income, perceived organizational commitment, and gender. In addition, we focus on the weekly interaction among these principals at leadership team meetings as they are expected to lead as an instructional leader and encouraged to collaborate on a regular basis with capacity to exchange leadership practice. Such frequent interaction around instructional leadership practice better suits the study context and thus would provide more relevant insight into principal collaboration as compared to the least frequent interaction such as yearly meetings.

Organizational commitment (OC) We used the organizational commitment scale developed by Bryk and Schneider [64] to suit the study context. The scale consists of five items on the same six-point scale. A sample item is, "I feel loyal to the district." Principal component analysis with promax rotation yielded a single factor solution that explained 74.8% of the variance with Cronbach's alpha of 0.90. The items and factor coefficient of organizational commitment are summarized in Table 2.

Table 2 Items, factor loadings, and reliability (Cronbach's alpha) of organizational commitment

Item	Factor loadings
Organizational commitment ($\alpha = .90$)	
1. I feel loyal to the district	.92
2. I usually look forward to each working day at the district	.90
3. I would recommend the district to parents seeking a place for their child	.86
4. I wouldn't want to work in any other districts	.83
5. I am committed to the programs and initiatives that enhance teaching and learning	.81

4 Data Analysis

We analyzed the relationships between individual principals' salary, their network position, and perceptions of organizational commitment in four steps. First, we used descriptive statistics including the network centrality measures to characterize the study sample. Second, as we are interested in two types of network relation (leadership advice and work recognition), we first employed quadratic assignment procedure (QAP)[1] to test the degree of correlation between leadership advice network and work recognition network. We then tested the correlation variables to explore the relationship between salary with network centrality measures and organizational commitment. Third, we created a set of sociograms to illustrate the relationships between principal network positions and their income and perception. Finally, this was followed by the factorial ANOVA analysis with a particular focus on the network position of principals in the two-plex network (LAWR) to test the association among gender, income, and organizational commitment with the principals' network position.

Social networks: two-plex ties and network position As we are interested in understanding the principals' network position formed by both advice-seeking and work recognition ties (two-plex relation), we examine their social position in such two-plex network in which a tie between any connected two principals represents the relation of both leadership advice *and* work recognition (LAWR). This two-plex approach, as opposed to uni-relation, allows the study to gain an in-depth understanding about not only the quantity of connections but also the strength of a tie that a principal has.

---

[1] Quadratic assignment procedure (QAP) is designed to test the statistical significance for social network data that is interdependence in nature. Unlike parametric statistical techniques, which assume observations that are analyzed are independent of one another, QAP is a nonparametric technique with no assumption of independence between observations. While using parametric statistics for social network data violates the assumption of independence, QAP is a suitable analytic strategy to test the statistical significance of social network data that are interdependent to one another. More information about QAP can be obtained from Hanneman and Riddle's (2005) tutorial.

In terms of identifying network position, we calculate key network centrality measures using the UCINET 6.0 software package [65] to quantify the degree of actor level connectedness that allows the study to interpret individual leaders' network position. We calculate closeness centrality and ego-reciprocity at the actor level to indicate the centricity of individual actors and the quality of ties. A recent study suggests the importance of closeness centrality in relation to educational leaders' advice-seeking behaviors in better facilitating the flow of resources across the network [8]. We thus calculate the closeness centrality particularly in an actor's incoming form, namely, *incloseness*, partly because advice-seeking behavior tends to be directional, as many of the network studies suggest [8]. When it comes to directionality of ties, it is important to consider the interpretation of ties and corresponding actor position. Since one's income may have to do with her/his recognizable status of performance among social peers [66], for our study it is more meaningful to investigate the level of recognized importance or popularity among network members. As such, incloseness serves a more suitable index to indicate an actor's quality network position than the outgoing ties of an actor. Incloseness of an actor n1 refers to the degree of normalized number of the shortest distances/paths required for the other actors to access actor n1 in a directed network. The incloseness of a principal in the present study corresponds to the proportion of the shortest path of incoming ties received from other principals. The normalized incloseness range is between 0 and 1. The greater the incloseness index, the more closely connected an actor is in relation to all other actors within a network. Thus, the information flow among actors of greater closeness tends to be more efficient, meaning it reaches larger segments of the network, than those of lower incloseness. In this manner, we may regard the incloseness of a principal as an index of the "quality" and "efficiency" of that principal's social status in accessing and disseminating leadership advice and work recognition.

In addition, in a team-based environment in which there is less hierarchical structure in roles and responsibilities, reciprocated ties with regard to mutual appreciation of others' efforts foster collaboration [22, 67, 68]. As such, network measure of *ego-reciprocity* serves as another important index to understand individual principals' position in the LAWR network. The ego-reciprocity of a principal refers to the proportion of reciprocated ties the principal has with others over all possible reciprocated ties across the network. Both actor incloseness and ego-reciprocity are presented in a form of percentage.

Descriptives and correlations We calculated descriptive statistics for the scale that measures organizational commitment and for network centralities. The QAP correlation between LA and WR networks indicates a statistically significant but weak correlation ($r = 0.39$, $p < 0.001$), suggesting that LA and WR each measures its idiosyncratic relationship and thus would yield added value in the two-plex network. We therefore examined the principal network position (incloseness and ego-reciprocity) in the LAWR two-plex network in the correlation analysis.

Factorial ANOVA analysis As we attempt to explore the association between the study variables and the network position of principals, we conducted factorial ANOVA analysis to test the effects of gender, income, and perceived organizational commitment on the principals' position (i.e., incloseness and ego-reciprocity) in the LAWR network. We primarily focus on gender and income as key demographic variables instead of experience, in that individual principals' income is highly correlated with their years of being an educator, in administration, and in current position. As such, we may consider income as a proxy measure of individual principals' human capital as defined by one's years of experience.

5 Results

5.1 Descriptive Statistics

Our descriptive statistics (see Table 3) suggest a number of patterns of the principal position in the LARW network. In the LAWR two-plex network, individual principals were nominated by 5% of their principal colleagues as a leader who provides leadership advice *and* work recognition ($SD_{\text{LAWR incloseness}} = 0.91$). Additionally, an average of 17% of a principal's LAWR ties was reciprocated ($SD_{\text{LAWR reciprocity}} = 0.21$), meaning that individual principals tend to engage in 17% of their relationships in mutually exchanging leadership advice and work recognition with others. In terms of leaders' perception, overall, the principals perceived higher levels of organizational commitment on a six-point scale with mean scores of 5.02 ($SD = 0.94$).

5.2 Visualizing LAWR Network of Principals

We provide a few sample network sociograms to illustrate the relationship among study variables such as income, network position, and perception of organizational commitment. Figure 1 presents the LAWR network of relationships between

Table 3 Descriptive statistics

	Min.	Max.	Mean	SD
Demographics				
Years of being an educator	6.50	33.00	19.87	6.87
Years in administration	1.50	20.00	8.48	5.31
Annual base income	30829.85	127872.70	93873.34	26407.23
Perception				
Organizational commitment	2.20	6.00	5.02	0.94
Network position				
LAWR incloseness	3.45	6.01	5.09	0.91
LAWR ego-reciprocity	0.00	0.67	0.17	0.21

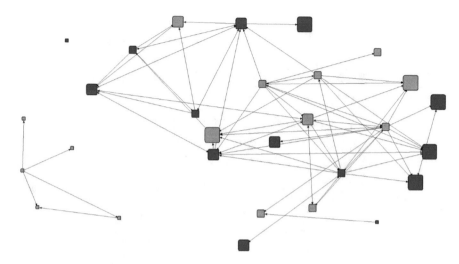

Fig. 1 LAWR network of relationship between pay and position. Note: Nodes are individual principals. Lines are the exchange of LAWR relationship. Nodes are colored by level of salary: orange = higher salary and navy = lower salary. Nodes are sized by level of incloseness: the larger the greater incloseness

individual principals' annual income and their network position in terms of degree of incloseness. Nodes in Fig. 1 are individual principals and are colored by level of income, with orange representing higher income and navy representing lower income. The cutoff point for high and low income is the mean annual income for principals in the district. Nodes are further sized by degree of incloseness—nodes with larger size representing the greater incloseness. Lines in Fig. 1 are directional, indicating the relationship between any two connected principals in seeking, providing, or reciprocating leadership advice and work recognition. Figure 1 shows that principals with higher annual income (orange nodes) are on average less likely to have closely connected network positions with their principal colleagues. This is particularly obvious for the cluster on the lower left corner in which there are five higher income principals who are of the lowest level of incloseness and disconnected from the main network.

Furthermore, we present another network to take into account the principals' perceptions of organizational commitment (OC) (Fig. 2). In Fig. 2, nodes and lines remain the same as Fig. 1, but the nodes are now grouped by level of income—higher income on the right and lower income on the left. Additionally, node color is now changed by level of OC perception—light green representing higher OC. Building on the preliminary findings from Fig. 1, Fig. 2 further indicates that the principals with higher income tend to be less well-connected but also perceive themselves as less committed to their school organization. This is reflected in the fact that only 2 out of 15 higher income principals perceive higher OC (13%), whereas 8 out of 14 principals from the lower income group perceive themselves as being committed to their school (57%).

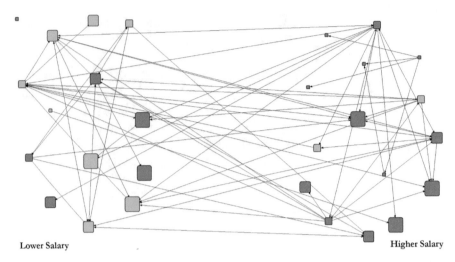

Lower Salary   Higher Salary

Fig. 2 LAWR network of relationship among pay, position, and organizational commitment. Note: Nodes are individual principals. Lines are the exchange of LAWR relationship. Nodes are grouped by level of salary: on the right = higher salary and on the left = lower salary. Nodes are colored by level of OC: light green = higher OC and gray = lower OC. Nodes are sized by level of incloseness: the larger the greater incloseness.

In sum, our preliminary findings from the network sociograms indicate potential differences in principals' annual income between different network positions and perceptions of OC. The following findings from our analysis will confirm these results.

5.3   Income, OC, Gender, and Network Position of Actor Incloseness

Figures 3 and 4 present the findings for actor incloseness while taking into account principals' income, perceived OC, and gender. Regardless of gender, Fig. 3 indicates a significant but negative difference in mean actor incloseness between higher income and lower income groups ($p < .05$). This suggests that principals with lower income are more closely connected to other principals in providing leadership advice and work recognition. Furthermore, while there was no significance in mean incloseness between different OC groups, for the lower OC group, we found a significant but negative difference in mean actor incloseness between higher income and lower income groups ($p < .05$). That is, for those principals who perceived themselves as being less committed to their schools, the difference in their network connectedness between income groups is significant with lower salary principals reporting higher incloseness.

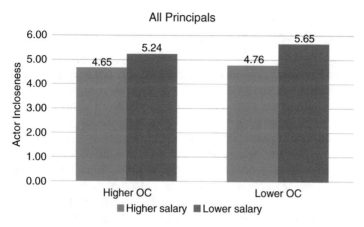

Fig. 3 Estimated marginal means of LAWR incloseness by salary and organizational commitment (OC). Note: The difference in mean actor incloseness between higher and lower salary is significant ($p < .05$). For lower OC, the difference in mean actor incloseness between higher salary and lower salary is significant ($p < .05$)

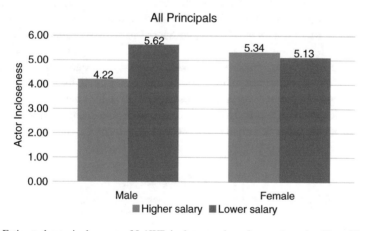

Fig. 4 Estimated marginal means of LAWR incloseness by salary and gender. Note: The difference in mean actor incloseness between higher and lower salary is significant ($p < .05$). For male, the difference in mean actor incloseness between higher salary and lower salary is significant ($p < .001$)

Regardless of perceived OC, the finding of difference in mean incloseness between different income groups is still the same as the finding in Fig. 3. However, when taking into account gender (Fig. 4), we found a significant but negative difference in the male group in terms of the mean actor incloseness between principals with higher and lower income ($p < .001$). That is, male principals' network position is subject to their income level—male principals with higher income are *less* well-connected with other principals in providing leadership advice and work recognition than those male principals with lower income.

5.4 Income, OC, Gender, and Network Position of Ego-Reciprocity

Figures 5 and 6 present the findings for ego-reciprocity and its relationships between principals' income, perceived OC, and gender. Regardless of perceived OC and gender, overall the difference in mean ego-reciprocity between higher income and lower income is significant and negative ($p < .05$). This suggests that principals with higher income, as compared to lower income group, are less likely to reciprocate the relationship of providing leadership advice and work recognition with other principals. This pattern of relationship between principals' network position in terms of ego-reciprocity and their income is similar to their closeness. As we compare different perceptions of OC with their network position (Fig. 5), we found a significance in the higher OC group between different income levels. That is, for those principals who perceived themselves as being committed to their school, their network position formed by reciprocal ties is subject to their income level. In other words, those principals with relatively lower income but higher in their perception of commitment are more likely to mutually exchange leadership advice and work recognition.

As we compare gender with their network position (Fig. 6), the difference in mean ego-reciprocity between male and female is significant ($p < .05$). Male principals (mean ego-reciprocity = 0.27), as compared to female principals (mean ego-reciprocity = 0.11), tend to have larger proportion of reciprocal ties with other principals in exchanging leadership advice and work recognition. Furthermore, when taken into account their income levels, this male principals' network position in terms of ego-reciprocity varies significantly ($p < .01$). In other words, those male principals, despite with relatively lower income, were able to engage in more mutual relationships with other principals than other male principals with higher income.

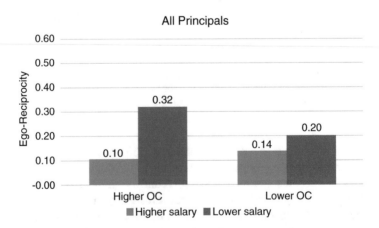

Fig. 5 Estimated marginal means of LAWR ego-reciprocity by salary and organizational commitment (OC). Note: The difference in mean ego-reciprocity between higher and lower salary is significant ($p < .05$). For higher OC, the difference in mean ego-reciprocity between higher salary and lower salary is significant ($p < .05$)

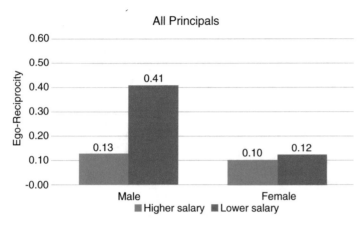

Fig. 6 Estimated marginal means of LAWR ego-reciprocity by salary and gender. Note: The difference in mean ego-reciprocity between higher and lower salary is significant ($p < .05$). The difference in mean ego-reciprocity between male and female is significant ($p < .05$). For male, the difference in mean ego-reciprocity between higher salary and lower salary is significant ($p < .01$)

5.5   Higher Income, OC, Gender, and Network Position

As we split the principals by income level, we found a few significant findings that are noteworthy. Figures 7 and 8 present the findings for high-income group of principals. For the high-income group, OC did not make any significant difference in influencing principals' network position. However, female principals with higher income are more well-connected with other principals than male principals in terms of incloseness connectivity (Fig. 7).

5.6   Lower Income, OC, Gender, and Network Position

While OC was not significantly associated with principals' network position in the higher income group, it was one key element in differentiating whether a principal with lower income level engages in more mutual exchange activity, especially for male principals (Figs. 9 and 10). For lower income group, degree of connectedness was not significant for gender and OC (Fig. 9). However, when it comes to mutual exchange (Fig. 10), we found that on average male principals (ego-reciprocity lower income male = 0.37) significantly engaged in more reciprocal activities than female principals (ego-reciprocity lower income female = 0.15). Of these male principals, despite received relatively lower income, those with higher OC (ego-reciprocity lower income, higher OC = 0.56) are significantly exchanging approximately three times as many mutual ties as those with lower OC (ego-reciprocity lower income, lower OC = 0.19). This suggests that being committed to school organizations is critical for principals to make a significant effort for sharing, exchanging work-related advice for collaboration without a greater monetary incentive.

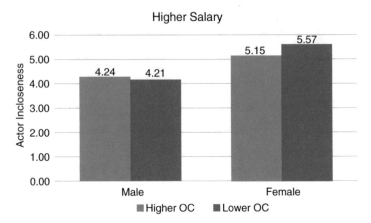

Fig. 7 Estimated marginal means of LAWR *incloseness* by gender and OC for higher salary. Note: There is no significance in mean actor incloseness between higher OC and lower OC. The difference in mean actor incloseness between male and female is significant ($p < .05$)

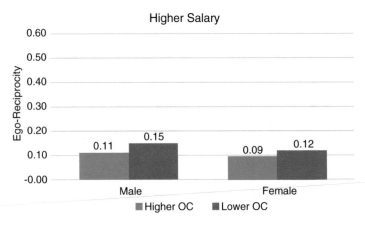

Fig. 8 Estimated marginal means of LAWR *ego-reciprocity* by gender and OC for higher salary. Note: There is no significance in mean ego-reciprocity between higher OC and lower OC

6   Discussion and Conclusion

In this chapter we examined the relationship among principals' network position, organizational commitment, and gender in a midsized urban fringe public school district. The main findings indicate a number of differences in principals' network position by their annual income and gender. Specifically, we found that principals with lower salary received more nominations by other principals perceiving them as providers of leadership advice and work recognition and that male principals had

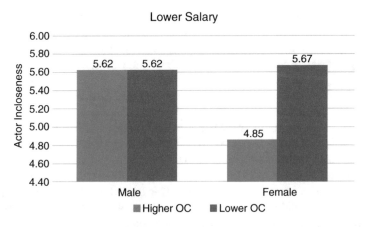

Fig. 9 Estimated marginal means of LAWR *incloseness* by gender and OC for lower salary. Note: There is no significance in mean actor incloseness between higher OC and lower OC. There is no significant difference in mean actor incloseness between male and female

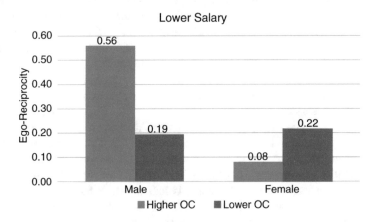

Fig. 10 Estimated marginal means of LAWR *ego-reciprocity* by gender and OC for lower salary. Note: There is no significance in mean ego-reciprocity between higher OC and lower OC. For gender, the difference in mean ego-reciprocity between male and female is significant ($p < .01$). For male, the difference in mean ego-reciprocity between higher OC and lower OC is significant ($p < .01$)

more mutual ties than female principals in exchanging/reciprocating leadership advice and work recognition. The former finding is counterintuitive to the existing understanding of salary and network position which suggests that central actors in an organization's power structure tend to have higher salary as they possess a vantage point of bargaining power for wage and benefits. Our finding indicates the opposite that the principals with lower salary are often regarded as providers of leadership advice and work recognition. As salary is highly correlated with seniority in the organization, as in our case, this finding also indicates that principals with fewer years of working in the position tend to be the central figures in the LAWR network. This may be due in part to the district's efforts around creating a nurturing, collaborative, and trusting culture of innovation. As a result, less experienced principals were not constrained by the district's existing routines and were able to exert their leadership practice and express novel ideas and as such were more likely to be regarded as the sources of advice and recognition.

Second, the latter finding confirms previous studies that indicate a dominating and central role of males in organizational network structures, especially for those employees with higher-rank positions (i.e., managers and executive leaders) and that same-gender ties tend to be reciprocal. Our study corroborates the reciprocal nature of such ties among male principals. This could be because male leaders are more inclined to build and sustain professional ties at work and for work purposes than female leaders, whose ties tend to be more social driven. As such, ties among male leaders would be more reciprocal. Building on this finding, our results further indicate this pattern of reciprocal ties—for male as opposed to female principals—is especially significant when they are in a lower salary pool. This reinforces the notion that male leaders are more likely to establish strong work ties despite the fact that their salary is relatively lower than their colleagues. This pattern of strong ties among male leaders might be even more conspicuous, particularly in organizations where men occupy the dominating leadership positions such as superintendent, managers, presidents, or coordinators, as in the study school district.

As for female leaders, contrary to the notion of the central role of male leaders, our findings indicate a central role of female principals when compared in the higher salary pool. Earlier studies indicate that in organizations where there are no or only a few women in high-ranking positions of power, women are more likely devalued [69] and often placed in lower organizational strata [70]. This is mainly because in such organizations women, more so than men, lack the resources, means, and motivation to challenge and change existing policies/criteria for success [ibid]. However, since a large portion of our study sample is female—as is the case in many other educational settings—women have more sufficient resources to act as influential actors in decision-making. The female principals with higher salary are likely to be regarded as more experienced and resourceful and occupy a more central position in proving leadership advice and work recognition. These gender findings in network positions signal another potential gap [71] we need to pay more attention to and that is to balance the gender role in the process of developing mutual collaboration and intellectual capital.

In terms of the role of organizational commitment in principals' network position, our main finding indicates that principals with lower salary but higher levels of organizational commitment engage in more work relationships with other principals for leadership advice and work recognition than principals with higher salary. This finding is especially salient among male principals. We found that male principals who perceive higher levels of organizational commitment despite having lower salary have approximately three times more reciprocal ties than the male principals with lower levels of commitment. These findings reinforce the important role of commitment in discerning employees' network position, confirming earlier studies that suggest that well-connected employees are more satisfied with and committed to work than the isolated ones [43]. In our study setting, reciprocal ties may be one of the professional norms the district aims to create, as reciprocity not only facilitates collaboration but also strengthens the development and sustaining of trusting relationships. As such, those "well-connected" principals may feel a strong sense of belonging as they mutually collaborate with one another, which in turn may result in greater commitment to their workplace, regardless of their salary level. Despite the fact that reciprocity may bring about positive influence on networking and collaboration, its downside should also be noted. Earlier studies indicate a negative effect of central network positions on employees' job satisfaction largely due to social liabilities [44, 72]. To prevent this disadvantage of reciprocity from happening, school districts may consider restructuring opportunities and resources in ways that include and engage "all" principals in leadership development such that all school leaders with various degrees of commitment would be connected, making the group as a whole heterogeneous in actors' perceptions of commitment and backgrounds in a hope to recreate a norm of cohesive commitment among all principals.

Our study makes a contribution to the social network literature by highlighting its relational capital in association of income, organizational commitment, and gender as well as implications to the work of school leaders. This chapter has implications on the dynamic of school network systems, in which organizational commitment is a critical issue in leading schools that are populated by diverse populations and full of educational reform challenges and uncertainties. Our work suggests that the crossing of income/gender boundaries as well as different levels of capital assets may be one way to enhance overall commitment to workplace and in a larger social system. While our work examined work-related network relation among principals, it may be worth investing in the social component in their interpersonal relationships, as the district aims to build a trusting and interconnected and collaborative environment. This speaks to the important role of informal/expressive network relations such as friendship, work energy, and social support in a work environment that is highly diverse. Investing in the capital of school leaders may help boost the overall commitment of employees, which in turn may minimize the differential role of salary and gender and may further sustain the long-term development of partnership among a leadership team across the school district.

References

1. Meyerson, E. M. (1994). Human capital, social capital and compensation: the relative contribution of social contacts to managers' incomes. *Acta Sociologica*, 37(4), 383–399.
2. Boxman, E. A., De Graaf, P. M., & Flap, H. D. (1991). The impact of social and human capital on the income attainment of Dutch managers. *Social networks*, *13*(1), 51–73.
3. Burt, R. S. (1992). *Structural holes*. Cambridge, MA: Harvard University Press.
4. Coleman, J. S. (1988). Social capital in the creation of human capital. *American Journal of Sociology*, 94, 95–120.

5. Granovetter, M. S. (1985). Economic action and social structure: the problem of embeddedness. *American Journal of Sociology*, 91, 481–510.
6. Nahapiet, J., & Ghoshal, S. (1998). Social capital, intellectual capital, and the organizational advantage. *Academy of management review*, *23*(2), 242–266.
7. Lin, N. (2009). *Social capital: A theory of social structure and action* (8th ed.). New York: Cambridge University Press.
8. Liou, Y.-H., & Daly, A. J. (2014). Closer to learning: Social networks, trust, and professional communities. *Journal of School Leadership*, 24(4), 753–795.
9. Scott, J. (2000). *Social network analysis*. London, UK: Sage Publications.
10. Lin, N. (1999). Building a network theory of social capital. *Connections*, *22*(1).
11. Lincoln, J., & Miller, J. (1979). Work and friendship ties in organizations: A comparative analysis of relational networks. *Administrative Science Quarterly*, 24, 181–198.
12. McPherson, J. M., & Smith-Lovin, L. (1987). Homophily in voluntary organizations: Status distance and the composition of face-to-face groups. *American sociological review*, 370–379.
13. Burt, R. S. (1976). Positions in networks. *Social Forces*, *55*, 93–122.
14. Brass, D. J. (1984). Being in the right place: A structural analysis of individual influence in an organization. *Administrative science quarterly*, 518–539.
15. Ronchetto Jr, J. R., Hutt, M. D., & Reingen, P. H. (1989). Embedded influence patterns in organizational buying systems. *The Journal of Marketing*, 51–62.
16. Friedkin, N. E. (1993). Structural bases of interpersonal influence in groups: A longitudinal case study. *American Sociological Review*, 861–872.
17. Bond, E. U., Walker, B. A., Hutt, M. D., & Reingen, P. H. (2004). Reputational effectiveness in cross-functional working relationships. *Journal of Product Innovation Management*, *21*(1), 44–60.
18. Kanter, R. M. (1988). When a thousand flowers bloom: structural, collective and social conditions for innovation in organizations. In Barry M. S., & L. L. Cummings (Eds), *Research in Organizational Behavior* (Vol. 10, pp. 169–211). Greenwich, CT: JAI Press.
19. Becker, G. S. (1992). *The economic way of looking at life*. Nobel Lecture, Chicago.
20. Becker, G. (1964). *Human capital: A theoretical and empirical analysis with special reference to education*. New York, NY: Columbia University.
21. Lindbeck, A. (1992). Macroeconomic theory and the labor market. *European Economic Review*, *36*(2–3), 209–235.
22. Bourdieu, P. (1986). The forms of capital. In J. C. Richardson (Ed.), *Handbook of theory and research for the sociology of education* (pp. 241–258). New York, NY: Greenwood Press.
23. Granovetter, M. S. (1973). The strength of weak ties. *American Journal of Sociology,* 78(1), 1360–1380.
24. Williamson, O. (1993). Calculativeness, trust, and economic organization. *The Journal of Law & Economics, 36*(1), 453–486.
25. Dasgupta, P. (2000). Trust as a commodity. In Gambetta, D. (Ed.), *Trust: Making and breaking cooperative relations* (pp. 49–72). Department of Sociology, University of Oxford.
26. Hackett, R. D., Bycio, P., & Hausdorf, P. A. (1994). Further assessments of Meyer and Allen's (1991) three-component model of organizational commitment. *Journal of applied psychology*, *79*(1), 15–23.
27. Mathieu, J. E., & Zajac, D. M. (1990). A review and meta-analysis of the antecedents, correlates, and consequences of organizational commitment. *Psychological bulletin*, *108*(2), 171.
28. Meyer, J. P., & Allen, N. J. (1997). *Commitment in the workplace: Theory, research, and Application*. Thousand Oaks, CA: Sage Publications.
29. Pittinsky, T. L., & Shih, M. J. (2004). Knowledge nomads: Organizational commitment and worker mobility in positive perspective. *American Behavioral Scientist*, *47*(6), 791–807.
30. Meyer, J. P., & Allen, N. J. (1991). A three-component conceptualization of organizational commitment. *Human Resource Management Review*, *1*(1), 61–89.
31. Mowday, R. T., Porter, L. W., & Steers, R. M. (1982). *Employee-organization linkages*. New York, NY: Academic Press.

32. Bashaw, R. E., & Grant, E. S. (1994). Exploring the distinctive nature of work commitments: Their relationships with personal characteristics, job performance, and propensity to leave. *Journal of Personal Selling & Sales Management, 14*(2), 41–56.
33. Dutton, J. E., & Ragins, B. R. E. (2007). *Exploring positive relationships at work: Building a theoretical and research foundation.* Mahwah, NJ: Lawrence Erlbaum Associates Publishers.
34. Katz, D., & Kahn, R. L. (1978). *The social psychology of organizations* (Vol. 2). New York, NY: Wiley.
35. Gittell, J. H., Cameron, K., Lim, S., & Rivas, V. (2006). Relationships, layoffs, and organizational resilience: Airline industry responses to September 11. *The Journal of Applied Behavioral Science, 42*(3), 300–329.
36. Dittes, J. E., & Kelley, H. H. (1956). Effects of different conditions of acceptance upon conformity to group norms. *The Journal of Abnormal and Social Psychology, 53*(1), 100.
37. Lee, J., & Kim, S. (2011). Exploring the role of social networks in affective organizational commitment: Network centrality, strength of ties, and structural holes. *The American Review of Public Administration, 41*(2), 205–223.
38. Brass, D. J., Galaskiewicz, J., Greve, H. R., & Tsai, W. (2004). Taking stock of networks and organizations: A multilevel perspective. *Academy of management journal, 47*(6), 795–817.
39. March, J. G., & Simon, H. A. (1993). Organizations revisited. *Industrial and Corporate Change, 2*(1), 299–316.
40. Mowday, R. T., Steers, R. M., & Porter, L. W. (1979). The measurement of organizational commitment. *Journal of vocational behavior, 14*(2), 224–247.
41. Mottaz, C. J. (1987). Age and work satisfaction. *Work and Occupations, 14*(3), 387–409.
42. Mottaz, C. J. (1989). An analysis of the relationship between attitudinal commitment and behavioral commitment. *The Sociological Quarterly, 30*(1), 143–158.
43. Roberts, K. H., & O'Reilly, C. A. (1979). Some correlations of communication roles in organizations. *Academy of management journal, 22*(1), 42–57.
44. Brass, D. J. (1981). Structural relationships, job characteristics, and worker satisfaction and performance. *Administrative science quarterly*, 331–348.
45. Krackhardt, D., & Porter, L. W. (1986). The snowball effect: Turnover embedded in communication networks. *Journal of Applied Psychology, 71*(1), 50.
46. Hartman, R. L., & Johnson, J. D. (1989). Social contagion and multiplexity communication networks as predictors of commitment and role ambiguity. *Human Communication Research, 15*(4), 523–548.
47. Wagner, W. G., Pfeffer, J., & O'Reilly III, C. A. (1984). Organizational demography and turnover in top-management group. *Administrative Science Quarterly*, 74–92.
48. McPherson, J. M., Popielarz, P. A., & Drobnic, S. (1992). Social networks and organizational dynamics. *American sociological review*, 153–170.
49. Hultin, M., & Szulkin, R. (1999). Wages and unequal access to organizational power: An empirical test of gender discrimination. Administrative Science Quarterly, 44(3), 453–472.
50. Rogers, E. M., & Kincaid, D. L. (1981). *Communication networks: toward a new paradigm for research.* New York, NY: Free Press
51. Brass, D. J. (1985). Men's and women's networks: A study of interaction patterns and influence in an organization. *Academy of Management journal, 28*(2), 327–343.
52. Ibarra, H. (1995). Race, opportunity, and diversity of social circles in managerial networks. *Academy of Management Journal, 38*(3), 673–703.
53. Moore, G. (1990). Structural determinants of men's and women's personal networks. *American Sociological Review,* 726–735.
54. Pugliesi, K., & Shook, S. L. (1998). Gender, ethnicity, and network characteristics: Variation in social support resources. *Sex Roles, 38*(3–4), 215–238.
55. Heyl, E. (1996). *Het docentennetwerk. Structuur en invloed van collegiale contacten binnen scholen.* Unpublished doctoral dissertation. University of Twente, The Netherlands.
56. McPherson, M., Smith-Lovin, L., & Cook, J. M. (2001). Birds of a feather: Homophily in social networks. *Annual review of sociology, 27*(1), 415–444.

57. Mehra, A., Kilduff, M., & Brass, D. J. (2001). The social networks of high and low self-monitors: Implications for workplace performance. *Administrative Science Quarterly, 46* (1), 121–146.
58. Fischer, C. S., & Oliker, S. J. (1983). A research note on friendship, gender, and the life cycle. *Social Forces, 62,* 124.
59. Marsden, P. V. (1987). Core discussion networks of Americans. *American sociological review,* 122–131.
60. Ibarra, H. (1992). Homophily and differential returns: sex differences in network structure and access in an advertising firm. *Administrative Sciences Quarterly, 37,* 422–47.
61. Leithwood, K., & Jantzi, D. (2008). Linking leadership to student learning: The role of collective efficacy. *Educational Administration Quarterly, 44*(4), 496–528.
62. Daly, A. J., Liou, Y., Tran, N. A., Cornelissen, F., & Park, V. (2014). The rise of neurotics: Social networks, leadership, and efficacy in district reform. *Educational Administration Quarterly, 50*(2), 233–278.
63. Cervone, H. F. (2008). Breaking out of Sacred Cow culture: The relationship of professional advice networks to receptivity to innovation in academic librarians. In Williams, D. & Golden, J., *Advances in library administration and organization* (Vol. 26, pp. 71–149).
64. Bryk, A. S., & Schneider, B. (2002). *Trust in schools: A core resource for school improvement.* New York, NY: Russell Sage Foundation.
65. Borgatti, S. P., Everett, M.G., & Freeman, L. C. (2002). *UCINET for Windows: Software for social network analysis.* Harvard, MA: Analytic Technologies.
66. Burt, R. S. (2000). The network structure of social capital. In R.I. Sutton & B.M. Staw (Eds.), *Research in Organizational Behaviour* (pp. 345–423). Greenwich, CT: JAI press.
67. Fearon, C., McLaughlin, H., & Morris, L. (2013). Conceptualising work engagement: An individual, collective and organisational efficacy perspective. *European journal of training and development, 37*(3), 244–256.
68. Wenger, E. (1998). *Communities of practice: Learning, meaning, and identity.* Cambridge, UK: University Press.
69. Pfeffer, J. (1989). A political perspective on careers: Interests, networks, and environments. In M. B. Arthur, D. T. Hall & B. S. Lawrence (Eds.), *Handbook of career theory* (pp. 380–396). New York, NY: Cambridge University Press.
70. Ely, R. J. (1995). The power in demography: Women's social constructions of gender identity at work. *Academy of Management journal, 38*(3), 589–634.
71. Turner, J. C., Sachdev, I., & Hogg, M. A. (1983). Social categorization, interpersonal attraction and group formation. *British Journal of Social Psychology, 22*(3), 227–239.
72. Labianca, G., & Brass, D. J. (2006). Exploring the social ledger: Negative relationships and negative asymmetry in social networks in organizations. *Academy of Management Review, 31*(3), 596–614.

Part IV
Tools and Techniques

Secondary Student Mentorship and Research in Complex Networks: Process and Effects

Catherine B. Cramer and Lori Sheetz

1 Background

Formal educational systems to date have not seriously considered how to engage students in understanding the explosion in big data and data-driven sciences that increasingly affect their lives, nor have they adequately prepared students for working and living in a data-driven society. Yet the twenty-first century STEM workforce needs specific skills in coping with and creating knowledge from large-scale streaming and multivariate data. The kinds of statistical, pattern-seeking, modeling, and probabilistic attributes inherent in so-called big data demand a thorough fluency in both exploratory and inductive skills to identify patterns and characterize their behavior across a wide range of differing environments and processes. Network science has emerged as a promising way to address these data-intensive, real-world problems.

Network science has its roots in graph theory, the mathematical/statistical study of connected systems as points (nodes or vertices) connected by lines (edges or links), which has been in use for over 250 years [1]. Graph theory was further developed for use in social network analysis (using graph theory to analyze the relationships in systems of social interaction), as interest in the applications of gestaltism beyond psychology emerged early in the twentieth century [2] and has been widely used since the 1950s [3]. Modern network science emerged toward the end of the twentieth century with the advent of powerful microcomputers, making it possible to apply statistical analysis and systems theory to large, complex datasets, seeking patterns and leveraging them for improved knowledge management and discovery

C. B. Cramer (✉)
Data Science Institute, Columbia University, New York, NY, USA

L. Sheetz
Center for Leadership and Diversity in STEM, Department of Mathematical Sciences,
U.S. Military Academy at West Point, West Point, NY, USA

[4]. Network science today is being used to understand everything from the human brain, to the origins of cancer, to the growth of cities, to our impact on the environment [5–7]. As the field of network science has grown and matured, so has its potential for improving understanding of complex natural and human-made phenomena in a number of domains. Its potential beneficiaries include not only researchers but also teachers, policy makers, businesspeople, and the general public as they cope with big data and complexity on a daily basis. As early as the 1970s, Jay Forrester identified the need for integrating complex connected and dynamical systems into K-12 education as an important life skill [8]. And there have been an increasing number of calls to include network theory in some form in K-12 education over the past 20 years [9–14].

However, over the last 25 years, there have been just a few disparate efforts to use network science, or graph theory, as a tool for instruction, include it in K-12 curriculum as a tool for students to use, or bring network science concepts and tools to a broad audience. Some examples include:

- The Mega Mathematics Project [15] developed by Los Alamos National Laboratory which includes a significant curriculum on graph theory but has not been widely adopted.
- The New York Hall of Science developed Connections: The Nature of Networks in 2004, the first public museum exhibition devoted exclusively to network science and emergence [16]. Although attended by millions of museum visitors, many of whom came away with a deeper understanding of the science and technology of connected systems, many visitors also struggled to develop a coherent understanding of the breadth and importance of network science and the exhibition closed in 2015 [17].
- The MapStats Curriculum developed by Rice University [12] attempted to introduce the statistical nature of "traveling salesman" type problems for middle school-aged students but was not adopted in schools;
- Connected: The Power of Six Degrees, a documentary on network science, aired briefly on the Science Channel in 2008; and
- The traveling exhibition Geometry Playground developed in 2009 by the Exploratorium includes activities on graph theory and networks, but even though visitor studies indicated that it improved attitudes and confidence toward geometry knowledge, it did not improve visitors' spatial reasoning and traveled to only a handful of museums [18].

These and other projects and programs intended to bring graph theory and network analysis to students and broader audiences have been slow to gain momentum, have not scaled well, are poorly coordinated, and are without a generalizable scope or structure.

The challenges revealed by these efforts indicated a need to rethink approaches to advanced learning of network science, making it more relevant to learners and providing opportunities to develop expertise in computational tools and network thinking. Network science can provide a pathway for students to learn about traditional topics across many disciplines, including social studies, science, computer science, and technology. Many of the problems explored through a network science approach are in the everyday experience of students and network conventions are part of our common vernacular, such as the network flow of air traffic, interconnectivity of coupled networks in political and social systems, and human networks as seen through technology activities such as Facebook, Twitter, and online social gaming. Computer platforms and tools for statistical analysis necessary for network analysis have become accessible to all and are increasingly intuitive and powerful to use. Yet exposure to these data-driven science skills has been unavailable to most students, and experiences working under researchers' or mentors' guidance are often most inaccessible to students underrepresented in STEM careers.

2   A New Approach: NetSci High

In 2010, a new effort called "NetSci High" was launched with a goal to bring network science into K-12 and informal learning settings. It began as the result of brainstorming at a network science meeting at the MIT Media Lab in which it was theorized that a pathway into accelerating student learning through deep engagement [19–21], and meaningful team research [22–26] could result in successful and enduring impact on heterogeneous groups of high school students. Goals for student learning were identified as follows:

1. Improve computational and statistical thinking and stimulate interest in computer programming and computational scientific methods by providing students and teachers with opportunities to create and analyze network models for real-world problems through a mentoring and training program and team research.
2. Increase students' potential for success in STEM in a technical career or college through applied problem solving across the curriculum with tested units of instruction that elucidate complex STEM topics and provide new applied approaches for critical thinking in STEM.
3. Prepare learners for twenty-first century science and engineering careers through the use of data-driven science literacy skills and motivate them to elucidate social and scientific problems relevant to the disciplines and their lives.
4. Help learners develop the following set of basic skills that are crucially needed to succeed in the twenty-first century data-driven work environment:

   - *Ability to synthesize*, seek, and analyze patterns in large-scale data systems.
   - *Improve facility with data visualization*, filtering, federating, and seeking patterns in complex data.
   - *Understand the changing role of models*, higher-order thinking, emphasizing exploratory skills to identify and characterize behavior of patterns in differing environments.
   - *Use network science and statistical approaches* to break down traditional silos to compare and contrast processes across domains.
   - *Build data fluency* to be able to identify, clean, parse, process, and apply appropriate analysis skills to large quantities of data.

- *Gain facility with data mining and manipulation* with increasingly semantic and statistical approaches, superseding logic models for searching and comparing data.
- *Understand the role of data sharing,* collaboration, interoperability of tools and data types, along with skills in using collaborative tools and methods to maximize data discovery.

Using a small amount of start-up support from NSF's Cyber-Enabled Discovery and Innovation fund, student/teacher teams were matched to network science research labs in the Northeast for year-long research projects, culminating in participation in the 2011 International School and Conference of Network Science (NetSci) at the Central European University in Budapest, Hungary. Because of the small amount of funding, only two of the seven student teams could be supported for travel, so the research projects were judged by a scientific committee made up of renowned network researchers, and the two winning project teams (from Thomas Edison and Flushing International High Schools in New York City) were sent to the conference to defend their work. All of the seven posters resulting from the research were presented at the poster sessions of the conference. The posters were also displayed at the 2011 Eighth International Conference on Complex Networks in Boston. The second year of NetSci High (2011–2012) offered scholarships to high school student teams. Two student teams participated from the Binghamton, NY, area, and the teams received a scholarship to attend the NetSci 2012 conference in Evanston, IL, and present their posters.

3 Network Science for the Next Generation

These earlier efforts resulted in positive outcomes for the participating students, enough so to encourage the organizers to scale up the project through funding from the National Science Foundation. The partnership expanded to include researchers from Boston University, and in 2012 the team was awarded a 3-year Innovative Technology Experiences for Students and Teachers (ITEST) grant from the National Science Foundation. The successful proposal – Collaborative Research: Network Science for the Next Generation – had a primary goal of providing a pathway for underserved high school students to engage in cutting-edge network science research, making accessible the power of network modeling and analysis. It was designed to provide opportunity for developing rigorous skills-based curricula and resources that utilize the rapidly growing science of complex networks, offering underserved students a context in which to learn computational and analytical skills for network-oriented data analysis. Importantly, it also gave the students a setting in which they could use these skills in original and independent research, leading to breakthroughs in solving large-scale, real-world problems and opening doors for future educational and career opportunities. The projects entailed intensive training and support of high school student teams and an academic year of research with cooperating university-based network science research labs; the labs' participation is facilitated by a graduate student who learns valuable mentorship skills as part of the experience. Because the network science field is relatively new, much of this research is novel, with practical implications.

Network Science for the Next Generation was designed as a regional educational outreach program. Over the course of the 3-year grant period (2012–2015), the project team formed close relationships with teachers and administrators from Title 1 high schools in both urban and rural areas in the Northeastern United States. ("Title 1" refers to a US Federal Department of Education program that provides financial assistance to schools with high numbers or high percentages of children from low-income families to help ensure that all children meet challenging state academic standards). By engaging Title 1 schools, the project adhered to its goal of providing opportunities to the students most likely to be left out of STEM career opportunities. Teachers recruited students for the project who were willing to make a commitment to a year-long project; however, in most cases, the students were not necessarily considered college-bound. The project team also recruited graduate students from research labs focused on network science applications, who would mentor the students throughout the school year.

These small teams–each made up of a teacher, 4 to 6 students, plus the grad student–were all required to attend an intensive 2-week summer workshop at which student teams, their teachers, and graduate student mentors are immersed in network science concepts; learn programming languages such as Python, NetLogo, and JavaScript; apply network analysis tools such as NetworkX and Gephi; attend hands-on workshops and talks from top network science researchers; and collaboratively brainstorm about research questions that will form the basis of the year's research projects. During the school year, the teams met regularly during and after school and submitted weekly progress reports and went on field visits to university labs, corporate headquarters, and science museums.

The culmination of each year was the presentation of the students' work at a major research conference; their posters were presented at the International School and Conference of Network Science in 2013, 2014, 2015, and 2016, and in several instances, the students themselves were present to defend their posters. In 2015 the organizers received a supplemental grant from NSF in order to bring all of the project students and teachers to the International Conference on Complex Networks (CompleNet). Each of these poster presentations experiences gave the students the opportunity to meet and talk with renowned researchers. In addition, one of the student projects–Spread of Academic Success in a High School Social Network–was published in PLOS One [27] (see Appendix for a complete list of student research projects).

4   Outcomes

The project bridged information technology practice and advances in network science research and provided career and technical education opportunities for young people underrepresented in data-driven STEM. The project team developed and implemented a rich, experiential, research-based program for 120 disadvantaged

high school students, 30 science research graduate student mentors, and 20 high
school STEM teacher mentors throughout New York State and Boston,
Massachusetts. It provided participants with opportunities to engage with top network science researchers, to be exposed to university and research lab settings and
potential career paths, to receive intensive skill-building and training, and most
importantly, to make real contributions to the field through original research. For
many of these students, this was a life-altering experience. The project also gave rise
to several Network Science and Education ancillary efforts, described below, which
are ongoing.

5   Evaluation

External evaluation of Network Science for the Next Generation (formative and
summative) was conducted by Davis Square Research Associates [28, 29] (Figs. 1
and 2). Evaluation looked at a range of project goals such as proximal (e.g., increasing high school student competencies in computing and improving student attitudes toward computing) as well as highly distal (e.g., preparing students for
twenty-first century science) learning outcomes, the emphasis on doing real-world
research into relevant and ambiguous problems through technologically infused
and highly collaborative projects, and cognitive goals such as analyzing, synthesizing, and visualizing quantitative data and understanding modeling, network statistics, etc. Evaluation also looked at outcomes of improving participant attitudes
toward the study of networks, self-efficacy, and the strengthening intentions of
continuing to pursue further involvement in college studies and careers. Guiding
research questions included:

- What is the effect of participation on the students' understanding of networks
  and their skills at analyzing networks?

Fig. 1 Pre-Post density
plots

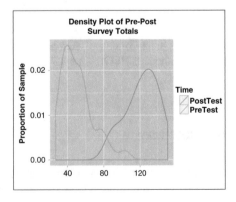

Fig. 2 Post-Test by
Pre-Test

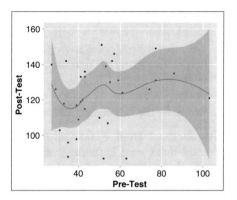

• How does participation affect the students' attitudes toward their own self-efficacy in computational approaches to networks and their commitments to continue to study networks in college and beyond?

Evaluation methods included gathering qualitative and quantitative data through focus groups, surveys, and SNA analyses. A significant challenge to the evaluation was the innovative nature of the project; since there was no or very little comparable activity, it was not possible to analyze any meaningful comparable groups.

Summative evaluation was intended to gather information on four pre-post constructs: knowledge of networks, intentions to continue studying networks, attitudes toward network science in general, and frequency of communications about networks [29]. Responding students reported large pre-post gains, especially in the areas of knowledge of network science and attitudes toward the study of networks. The effect sizes are very large, though these may be somewhat inflated due to a social desirability bias associable with the retrospective pretest design of the survey. None of the four constructs (knowledge, intentions, attitudes, communications) showed gain scores that were significantly clustered around the mean ($p < 0.05$, Kolmogorov-Smirnov [30]), indicating that some students gained more and others gained less, as could perhaps be inferred from the large standard deviations noted. The same observation held true for the aggregated pre-post gain scores.

The pretest scores were not predictive of the posttest scores. This is a good finding in that it suggests that the project worked comparably well with students no matter what their levels of preparation at the beginning of their participation. The following density plots show the very stark shift from the pretest to the posttest moments. The gray areas express the confidence interval.

Participating teachers were surveyed using open-ended questions about student network learning, student skills development, and a sustained student interest in careers. For each of the questions, teachers were asked to reflect on their own learning, their own network skills development, and the potential for sustained effects on their own practice. Responding teachers indicated that:

• They tended to cite the value of the more theoretical learning, with a special emphasis on the innovative qualities of the project.

- They were highly appreciative of students' new technical skills, with these skills clearly seen as having a distinct educational value.
- They saw the project as having an enormous educational value, with the network perspective likely to continue to influence the participating educators' approach to certain content. The previous finding that many students intended to continue to study networks, or to incorporate networks in their future studies, was confirmed by the responding teachers.

Overall, the external evaluation found that the project was consistently effective over the three years in reaching its most basic goals of improving students' understanding of networks, their self-efficacy with regard to the study of networks, the attribution of value regarding networks, and their intentions to continue to study networks. Key findings include:

- Participating students reported significant cognitive gains in their awareness of networks and their skills at using computers to analyze networks.
- Participating students reported significant gains in their attitudes toward the value of studying networks and their confidence in being successful at such studies.
- Both participating students and teachers reported they intended to continue to use network approaches.

Significant project outcomes revealed:

- Original student and teacher research projects are not only possible but form an essential incentive and commitment for participants to remain engaged in and to bring projects to completion.
- It is possible to train a broad spectrum of students and teachers in enough computer programming (e.g., Python or R) to use sophisticated network analysis tools within programing environments.
- A supportive community and consistent mentorship are essential to success.
- Teachers can assume an active leadership role in mentoring students who are engaged in network science research provided adequate supports are in place.
- Students and teachers are remarkably innovative in terms of how they develop and pursue project-based learning approaches in network science.

Students reported high levels of engagement in projects that used real-world, active learning in highly collaborative teams. Overcoming the social barriers to collaboration and the technical and logistical challenges all proved to be of great worth to the participating students. The role of the faculty and graduate student mentors was also of great value, providing prompt and clear feedback as the projects developed. This combination of meaningful challenge with attentive and thoughtful guidance is of the greatest educational value. It is extremely effective for students in incentivizing and achieving success in mastering and applying data-driven STEM skills to real problem solving and is an empowering and engaging pathway into data and computational literacy and computer programming skills.

6 Conclusions

Over the past 8 years of NetSci High, it has become evident that a successful program of learning about the science of complex networks is achievable for heterogeneous groups of high school students from underrepresented groups, provided there is the right support "ecosystem." This ecosystem includes:

- Deep engagement in learning about network science and applying network science tools and techniques in a variety of ways, including training in enough computer programming (e.g., Python or R) to use sophisticated network analysis tools
- Highly collaborative student-teacher-mentor teams that persist throughout the life of the project
- Opportunity for students to innovate in terms of how they develop and pursue project-based learning approaches in network science, make mistakes, and explore a variety of avenues for creativity
- Original team research projects that form an essential incentive and commitment for participants to remain engaged and to bring projects to completion
- A supportive community and consistent mentorship, including an active leadership role for teachers who are allowed enough time to mentor students and are able to maintain contact with researchers

It became immediately apparent during the first year of Network Science for the Next Generation that participating teachers were most effective if they had, at a minimum, the same level of training as their students before being called upon to mentor students: they wanted to be active mentors, rather than co-learners with their students. As the project progressed, interactions with teachers became deeper and more meaningful, and teachers took on new roles, pursued their own interests, and took ownership of network science approaches. Original student and teacher research projects are not only possible but form an essential incentive and commitment for participants to remain engaged in and to bring projects to completion. A path to scalability began to emerge.

NetSci High has made significant educational impacts on regional high school students and teachers and is also prompting strong social commitments from the network science community as a whole [31, 32]. It aims to address the challenge of transforming the way we educate our citizens to keep pace with not only the amount of data we collect but to appreciate how network science identifies, clarifies, and solves complex twenty-first century challenges in the environment, medicine, agriculture urbanization, social justice, and human well-being. NetSci High provides a pathway to integrate science research and programming skills for high school students who would not otherwise have these opportunities. Additionally, it encourages high school teacher mentors to broaden their STEM understanding and informs their current teaching in terms of content and practice. Success in NetSci High elicits students to persist in interest in STEM and computational fields and attain ambitious academic goals.

7 Other NetSci Community Efforts

The work described above has resulted in the growth of an active Network Science in Education global community, also known as NetSciEd. Several ongoing streams of collaborative work have resulted.

7.1 NetSciEd Satellites

NetSci High organizers are dedicated to supporting the growth of network science education efforts. A key venue for support takes place at the annual International School and Conference on Network Science (NetSci), where the Network Science and Education (NetSciEd) satellite symposium has taken place since 2012 [33]. NetSciEd provides an opportunity for the growing community of network scientists interested in education to come together to exchange ideas, talk about their projects and programs, demonstrate and get feedback on new tools, and discuss curriculum models.

7.2 Network Literacy

Attaining a basic understanding of networks has become a necessary form of literacy for all people living in today's society and for young people in particular. At the 2014 International School and Conference on Network Science, a year-long process was initiated to develop an educational resource that concisely summarizes essential concepts about networks that can be used by anyone of school age and older. The process involved several brainstorming sessions on one key question: "What should every person living in the twenty-first century know about networks by the time he/she finishes secondary education?" These brainstorming sessions reached diverse participants, which included professional researchers in network science, educators, and high school students. The generated ideas were connected by the students to construct a concept network. We examined community structure in the concept network to group ideas into a set of important concepts, which, through extensive discussion, we further refined into seven essential concepts [34]. The students played a major role in this developmental process by providing insights and perspectives that were often unrecognized by researchers and educators. The final result, "Network Literacy: Essential Concepts and Core Ideas," is now available as a booklet and has been translated into 20 languages [35]. (See Appendix for complete text.)

7.3   Learning Settings

Among the goals of NetSci High was to have teachers integrate network science into instruction. Concurrent with the NetSci High student training and research project is a multiphase, multiyear approach to professional development with formal and informal educators, including efforts to facilitate the development of practices and resources with educators and promote network thinking among K-12 students, teachers, and administrators through developing curriculum materials, lesson plans, and practical learning resources for K-12 classrooms across all domains of knowledge; providing rigorous professional development opportunities for both formal (school) and informal (cultural institutions, camps, after-school, and other community-based programs) educators; increasing the awareness of the demand for network science education among researchers; and developing a path for mapping *Network Literacy Essential Concepts and Core Ideas* to learning standards such as Next Generation Science Standards (NGSS) [36]. And, most recently, one of the NetSci High participating school districts includes a class on Network Literacy in its official after-school course catalogue, taught by a participating NetSci High veteran educator.

Acknowledgments The authors would like to acknowledge the National Science Foundation (BCS Award #1027752 and DRL Award #1139478) for supporting this important work. Any opinions, findings, and conclusions or recommendations expressed in this paper are those of the authors and do not necessarily reflect the views of the National Science Foundation.

Appendices

NetSci High Student Research Projects 2010–2015

2010–2011

- A Comparative Study on the Social Networks of Fictional Characters
- Academic Achievement and Personal Satisfaction in High School Social Networks
- Does Facebook Friendship Reflect Real Friendship?
- Inter-Species Protein-Protein Interaction Network Reveals Protein Interfaces for Conserved Function
- The Hierarchy of Endothelial Cell Phenotypes
- Preaching To The Choir? Using Social Networks to Measure the Success of a Message
- Identification of mRNA Target Sites for siRNA Mediated VAMP Protein Knockdown in *Rattus norvegicus*

2011–2012

- A Possible Spread of Academic Success in a High School Social Network: A Two-Year Study
- Research on Social Network Analysis from a Younger Generation

2012–2013

- Interactive Simulations and Games for Teaching about Networks
- Mapping Protein Networks in Three Dimensions
- Main and North Campus: Are We Really Connected?
- High School Communication: Electronic or Face-to-Face?
- An Analysis of the Networks of Product Creation and Trading in the Virtual Economy of Team Fortress 2

2013–2014

- A Network Analysis of Foreign Aid Based on Bias of Political Ideologies
- Comparing Two Human Disease Networks: Gene-Based and System-Based Perspectives
- How Does One Become Successful on Reddit.com?
- Influence at the 1787 Constitutional Convention
- Quantifying Similarity of Benign and Oncogenic Viral Proteins Using Amino Acid Sequence
- Quantification of Character and Plot in Contemporary Fiction
- RedNet: A Different Perspective of Reddit
- Tracking Tweets for the Superbowl

2014–2015

- Network Analysis of Microgravity-Influenced Genes in *Salmonella enterica serovar typhimurium*
- Connecting Radon Levels to Cancer rates in California Counties: A Network Approach
- National Football League Network
- Drug combinations and adverse side effects
- Comparing post-secondary institutions across the United States
- Relationships Between Musculoskeletal System and High School Sports Injuries
- Similarities Found in Neurological Disorders Based on Mutated Genes and Drug Molecules
- The Relationships of International Superpowers
- Protein Association and Nucleotide Similarities Among Human Alpha-Papillomaviruses

Network Literacy: Essential Concepts and Core Ideas

1. Networks are everywhere.

   - The concept of networks is a broad, general idea about how things are connected and working together. Networks are present in every aspect of life.
   - There are networks that form the technical infrastructure of our society (e.g., communication systems, the Internet, the electric grid, the water supply).
   - There are networks of people, e.g., families and friends, email/text exchanges, Facebook/Twitter/Instagram, and professional groups.
   - There are economic networks, e.g., financial transactions, corporate partnerships, and international trades.
   - There are biological/ecological networks, e.g., food webs, gene/protein interactions, neural networks, and spread of diseases.
   - There are cultural networks, e.g., language, literature, art, history, and religion.
   - Networks can exist at various spatial and/or temporal scales.

2. Networks describe how things connect and interact.

   - There is a subfield of mathematics that applies to networks. It is called graph theory. A graph in mathematics means a network.
   - Connections are called links, edges, and ties. Things that are connected are called nodes, vertices, and actors.
   - Connections can be undirected (symmetric) or directed (asymmetric).
   - The number of connections a node has is called a degree of that node.
   - In some networks, you can find a small number of nodes that have much larger degrees than others. They are called hubs.
   - A sequence of links that leads you from one node, through other nodes, to another node is called a path.
   - In some networks, you can find a group of nodes that are well connected to each other. They are often called clusters, cliques, and communities.

3. Networks help reveal patterns.

   - You can represent something as a network by describing what its parts are and how they are connected to each other. Such network representation is a very powerful way to study its properties.
   - Some of the properties in a network that you can study are:

     - How the degrees are distributed across nodes.
     - Which parts or connections are the most important ones.
     - Strengths and/or weaknesses of the network.
     - If there is any substructure or hierarchy.
     - How many hops, on average, are needed to move from one node to another within the network.
     - Using these findings, you may be able to make predictions.

4. Visualizations can provide an understanding of networks.

- Networks can be visualized in a number of different ways.
- You can draw a diagram of a network by connecting nodes with links.
- There are a variety of tools available for visualizing networks.
- Visualization of a network often helps to understand it and communicate the ideas to people in an intuitive, nontechnical way.
- Creative information design plays a very important role in making an effective visualization.
- It is important to be careful when interpreting and evaluating visualizations, because they may not tell the whole story about the networks.

5. Today's computer technology allows us to study real-world networks.

- Computer technology has dramatically enhanced our ability to study networks, especially large complex ones.
- There are many free software tools available for network visualization and analysis.
- Using personal computers, everyone can easily model, visualize, and analyze networks, not just scientists.
- Through the Internet, everyone has access to many interesting network data.
- Computers allow us to simulate hypothetical or virtual networks, as well as real ones.
- Learning basic computer literacy skills opens the door to infinite possibilities, e.g., file/folder operation, data entry, manipulation and modeling, information sharing and collaboration, and computer programming.

6. Networks help us compare a wide variety of systems.

- Various kinds of systems, once represented as networks, can be compared to see how similar or different they are.
- Certain network properties commonly appear in many seemingly unrelated systems. This implies that there may be some general network principles across disciplines.
- Other network properties are quite different from systems to systems. These properties can help classify networks in different families and understand them differently.
- Science has traditionally been conducted in separate disciplines. Networks can help go beyond disciplinary boundaries toward a more cross- or interdisciplinary understanding of the world.
- Networks can help transfer knowledge from one discipline to another to make a breakthrough.

7. Network structures can influence their dynamics and vice versa.

- Network structure means how parts are connected in a network.
- Network dynamics means how things change over time in a network.

- Network structures can influence their dynamics. Examples include the spread of diseases, behaviors or memes in a social network, and traffic patterns on the road network in a city.
- Network dynamics can influence their structures. Examples include the creation of new following links in social media and construction of new roads to address traffic jams.
- Network structures and dynamics often influence each other simultaneously.

Outreach Events

- NetSci High has facilitated sending a group of high school students and teachers from New York City to NetSci 2011 in Budapest, Hungary; a group from Endwell and Vestal, NY to NetSci 2012 in Evanston, IL; and a group from Vestal, NY to NetSci 2014 in Berkeley, CA. In all of these travels, the high school student teams presented their work at poster sessions. High school student research has also been published in peer-reviewed journals such as PLOS One (Blansky et al. 2013).
- Students and teachers from Newburgh Free Academy and Vestal High School presented posters at the IEEE ISEC conference at Princeton University in 2015 and 2016. NetSci High participating students and teachers have presented at the West Point Cadet Seminar on Network Science each year of NetSci High.
- In 2016 NetSci High students and teachers participated in the US Science and Engineering Festival in Washington, D.C., presenting several hands-on activities related to the Network Literacy Essential Concepts.

References

1. Euler, Leonhard (1736). Solutio problematis ad geometriam situs pertinentis. Commentarii Academiae Scientiarum Imperialis Petropolitanae 8. 128.
2. Moreno, J. (1934). *Who Shall Survive? A new Approach to the Problem of Human Interrelations.* Beacon House.
3. Barnes, J. (1954). Class and Committees in a Norwegian Island Parish. Human Relations 7. Thousand Oaks, CA: Sage Publications. 39.
4. Barabási A.-L., Albert R. (1999) Emergence of scaling in random networks, Science, Vol. 286, No. 5439. Washington, DC: American Association for the Advancement of Science. 509.
5. Barabási, A.-L. (2002). Linked: How everything is connected to everything else and what it means. New York: Plume.
6. Pastor-Satorras, R. and Vespignani, A. (2001) Epidemic spreading in scale-free networks, Physical Review Letters 86. Ridge NY: American Physical Society. 3200.
7. Lazer, D., Pentland, A., Adamic, L., Aral, S., Barabàsi, A-L., Brewer, D., Christakis, N., Contractor, N., Fowler, J., Gutmann, M., Jebara, T., King, G., Macy, M., Roy, D., and Van Alstyne, M. (2009). Computational social science. Science, 323. Washington, DC: American Association for the Advancement of Science. 721.

8. Forrester, Jay W (1976) Moving into the 21st Century: Dilemmas and Strategies for American Higher Education. Liberal Education, Vol. 62, No. 2. Washington, DC: Association of American Colleges & Universities. 158.
9. Hart, E., Maltas, J. and Rich, B. (1990) Teaching Discrete Mathematics in Grades 7–12. Mathematics Teacher Vol. 83, No. 5. Reston, VA: National Council of Teachers of Mathematics. 362.
10. Bollobas, B. (1998) Modern Graph Theory. Graduate Texts in Mathematics, Vol. 184. New York: Springer.
11. Wilson, S. and Rivera-Marrero, O. (2004) Graph Theory: A Topic for Helping Secondary Teachers. Paper presented at the annual meeting of the North American Chapter of the International Group for the Psychology of Mathematics Education.
12. Fuhrmann, S. , MacEachren, A. , Deberry, M. , Bosley, J., LaPorte Taylor, R. , Gahegan, M. & Downs, R. (2005) MapStats for Kids: Making Geographic and Statistical Facts Available to Children, Journal of Geography, 104:6, 233–241, DOI: https://doi.org/10.1080/00221340508978645.
13. Smithers, D. (2005) Graph Theory for the Secondary School Classroom. Electronic Theses and Dissertations. Paper 1015. http://dc.etsu.edu/etd/1015. Accessed 1/1/18.
14. Lessner, D. (2011), Graph Theory at Czech Grammar Schools, in WDS'11 Proceedings of Contributed Papers: Part I - Mathematics and Computer Sciences (eds. J. Safrankova and J. Pavlu), Prague, Matfyzpress. 78.
15. Casey, N., Fellows, M., and Hawylycz, M. (1993) This Is MEGA Mathematics: Stories And Activities For Mathematical Problem Solving And Communication, The Los Alamos Workbook. Los Alamos, NM: Los Alamos National Laboratory.
16. Siegel, E., and Uzzo, S. (2010) Connections: The Nature of Networks, Communicating Complex and Emerging Science. In Science Exhibitions, Communication and Evaluation. Edited by Anastasia Filippoupoliti. Edinburgh: MuseumsEtc.
17. Rothenberg, M. and Hart, J. (2006) Analysis of Visitor Experience in the Exhibition Connections: the Nature of Networks at the New York Hall of Science. Northampton, MA: People, Places & Design Research.
18. Dancu, T., Gutwill, J., and Sindorf, L. (2009) Geometry Playground Pathways Study. San Francisco, CA: Exploratorium.
19. Laibowitz, M. (2004) Meaning Density and Other Attributes of Deep Engagement MIT Media Lab report. Cambridge, MA: Massachusetts Institute of Technology.
20. Pugh, K., Linnenbrink-Garcia, L., Koskey, K., Stewart, V. and Manzey, C. (2010), Motivation, learning, and transformative experience: A study of deep engagement in science. Science Education, 94. New York: Wiley and Sons. 1.
21. Crick, R, (2012) Deep Engagement as a Complex System: Identity, Learning Power and Authentic Enquiry. in: Sandra L Christenson, S Reschly, C. Wylie (eds) *Handbook of Research on Student Engagement*. New York: Springer. 675.
22. Cohen, B., & Cohen, E. (1991). From groupwork among children to R&D teams: Interdependence, interaction and productivity. In E. Lawler, B. Markovsky, C. Ridgeway, & H. Walker (Eds.), Advances in group processes, Vol. 8. West Yorkshire: Emerald Publishing. 205.
23. Webb, N. (1992). Testing a theoretical model of student interaction and learning in small groups. In R. Hertz-Lazarowitz & N. Miller (Eds.), Interaction in cooperative groups: The theoretical anatomy of group learning. New York: Cambridge University Press. 102.
24. Garland, D., O' Connor, M., Wolfer, T., and Netting, F. (2006) Team-based Research Notes from the Field. Qualitative Social Work Volume: 5 No. 1. Thousand Oaks: Sage Publications. 93.
25. Hsu, L., Lee, K. and Lin, C, (2010) A comparison of individual and team research performance: A study of patents in III, Picmet 2010 Technology Management For Global Economic Growth, Phuket. 1.

26. National Research Council. 2015. Enhancing the Effectiveness of Team Science. Washington, DC: The National Academies Press. https://doi.org/10.17226/19007.
27. Blansky, D., Kavanaugh, C., Boothroyd, C., Benson, B., Gallagher, J., Endress, J., and Sayama, H. (2013). Spread of academic success in a high school social network. PLOS ONE, Vol. 8, No. 2. e55944.
28. Faux, R. (2014) Summer Workshop Evaluation Report. Boston, MA: Davis Square Research Associates.
29. Faux, R. (2015) Evaluation of the NetSci High ITEST Project: Summative Report. Boston, MA: Davis Square Research Associates.
30. Daniel, Wayne W. (1990). "Kolmogorov–Smirnov one-sample test". Applied Nonparametric Statistics (2nd ed.). Boston: PWS-Kent. pp. 319–330.
31. Harrington, H., Beguerisse-Díaz, M., Rombach, M., Keating L, and Porter, M. (2013) Teach network science to teenagers. Network Science, Vol.1, No. 2. Cambridge: Cambridge University Press. 226.
32. Sanchez, A., Brandle, C. (2014). More network science for teenagers. arXiv:1403.3618.
33. NetSciEd (2018) NetSciEd Symposia (https://sites.google.com/a/binghamton.edu/netscied/events/netscied-symposia) Accessed 1/1/18.
34. Cramer, C., Porter, M., Sayama, H., Sheetz, L. and Uzzo, S. (2015) What are essential concepts about networks? Journal of Complex Networks. Vol. 3 No. 4 Oxford: Oxford University Press.
35. NetSciEd (2015) Network Literacy: Essential Concepts and Core Ideas. (http://tinyurl.com/networkliteracy) Accessed 1/1/18.
36. NGSS Lead States. (2013). Next Generation Science Standards: For States, By States. Washington, DC: The National Academies Press.

The Imaginary Board of Directors Exercise: A Resource for Introducing Social Capital Theory and Practice

Brooke Foucault Welles

1 Introduction

People naturally find themselves in social groups, or *networks*, of friends, family, classmates, coworkers, and so on. These networks are associated with improvements in a variety of outcomes, including social and emotional well-being, educational and career achievement, and personal health [1–3]. Networks improve outcomes because of the resources, or *social capital*, that they can provide. Social capital refers to the resources that are available to us because we are embedded within networks of social relationships. It is generally described in two forms: *bonding* social capital, or the resources we obtain from our close relationships (family, friends, etc.), and *bridging* social capital, or the resources we obtain from more casual relationships (acquaintances, neighbors, etc.) [4, 5]. Social capital generates many kinds of resources, including material goods (e.g., the ability to borrow a ladder from a neighbor), information and advice, general social support and a sense of belonging, and the possibility to coordinate collective action [6].

However, simply being part of a network does not guarantee that someone will benefit from social capital [7]. Although social networks provide the *potential* to access resources, it is up to individuals to both *recognize* and *activate* those resources to benefit from that potential. Being able to successfully access networked resources depends critically on the ability to understand one's own complex web of interpersonal relationships and being able to accurately navigate that web of relationships to extract the right resources at the right time. These are challenging tasks, and people naturally vary in their abilities to accurately and effectively access resources in their own social networks [8], with some people navigating their social networks much more advantageously than others [9].

B. Foucault Welles (✉)
Northeastern University, Boston, MA, USA

© Springer International Publishing AG, part of Springer Nature 2018   159
C. B. Cramer et al. (eds.), *Network Science In Education*,
https://doi.org/10.1007/978-3-319-77237-0_10

Sometimes, people simply forget who is in their network. Research suggests that people routinely forget between 5% and 95% of their social ties [10]. The likelihood of forgetting social ties is a function of several things. The type of social tie matters – weak ties (those that yield bridging social capital) are substantially more likely to be forgotten than strong ties [ibid], and ties with people we see irregularly are more likely to be forgotten than those with people we see on a daily or weekly basis [11]. Network structure also plays a role. Dense networks, in which social ties tend also to be tied to one another, are easier to remember than sparser networks of the same size [12], and people who are centrally positioned in their social networks tend to recall ties more accurately than those who are on the periphery of the network [13]. Finally, individual psychology and social status affect recall accuracy. Certain personality characteristics including need for affiliation, extraversion, and self-monitoring are associated with better recall of social ties [ibid], while low social status is associated with poorer recall, especially under duress [14].

Low-income, minority individuals in the United States may find it especially difficult to access social capital effectively. Evidence suggests that these individuals have fewer overall resources within their networks and experience differential, lower returns on requests for resources due to social norms that discourage asking for and receiving favors from strategically positioned members of the community [15, 16]. As a result, social networks may reify existing race- and class-based social inequalities in the United States by exacerbating other limitations in access to social resources [17]. For these individuals, learning to successfully understand and leverage opportunities in their interpersonal relationship networks is especially critical.

2 The Imaginary Board of Directors Exercise

The imaginary board of directors exercise was developed to help low-income, minority high school students[1] learn about social capital and improve their chances of accessing it in service of a personal goal. Modeled after an exercise commonly used with MBA students [18], the hour-long activity invites students to create a network of people (the "board of directors") that can help them achieve academic and career goals. In the process, students learn basic concepts about social networks and social capital and identify individuals who can help them achieve their goals.

---

[1] "Minority," in this case, refers primarily to members of disadvantaged racial and ethnic minority groups in the United States. I typically run this exercise with Black and Latinx high school students in historically low-income portions of Boston, MA, and the surrounding suburbs. I have also successfully run the exercise with college freshmen, young professionals, and mixed-age audiences (children and adults). The examples given in the text are from my work with high school students but could easily be adapted as needed.

2.1 Introducing Social Networks and Social Capital

This exercise is designed for high school students (14–18 years old) with no prior knowledge of social networks or social capital. The exercise works best for groups of 20–40 students, although I have successfully implemented it in an auditorium of 200+ participants. It begins with a brief introduction to basic concepts of social networks and social capital. The introduction can take many forms depending on audience size, available resources, and presenter preferences. I typically lecture with notes on a whiteboard, but I have also used PowerPoint slides for larger audiences. Regardless of format, the presentation should include the following key points:

- Social networks are sets of people connected by relationships (also called "ties"). Everyone is embedded in many different social networks.
- There are many different types of social relationships, including strong ties and weak ties. Strong ties include people you are very close to, such as family members and close friends. Weak ties include people you know casually, such as classmates or people who live in your neighborhood.
- Strong ties and weak ties are both valuable because they provide us with different kinds of resources. We need both types of ties in our lives to achieve our goals. A common misconception is that strong ties are "good" and weak ties are "bad." It is critical to correct this misunderstanding.
- These resources are called *social capital* and are embedded inside networks. These resources are available to anyone who is in a network, and they can include a variety of things like material goods, advice, social support, and the possibility to organize groups to do things.

As you work through these points, it is helpful to invite students to reflect on what each of these key points means for them, personally. For instance, I ask students to list a few people they consider to be strong and weak ties. The answers to these questions can vary substantially for students with different backgrounds and life experiences. For example, some students feel very close to teachers, coaches, and other school staff, while others consider these people weak ties, or, in some cases, not part of their networks at all. Similarly, some students count extended family among their closest ties, while others have no contact with these people whatsoever. Having students reflect on these differences helps them to think about the diversity of ties they have in their networks and can also help me, as the discussion leader, choose relevant examples to illustrate key concepts in the overview.

2.2 Setting Goals

Following the introduction, I ask students to identify a personal goal that they cannot achieve on their own. The goal can be related to school, work, athletics, hobbies, or anything else that matters to them. It should be specific and something they hope to achieve in the relatively near future (1–3 years from the present is a good guideline). Occasionally, students will have difficulty imagining suitable goals. To help these students, it can be useful to invite other students in the group to share their goals. I also give examples of common goals–e.g., graduating high school with a certain Grade Point Average, finding a summer job, making a varsity sports team, and getting admitted to college–all of which are often mentioned by students in these groups. It is important not to discourage students from certain goals that may seem either unrealistic or overly simplistic. The exercise works equally well for ambitious goals (e.g., getting drafted into the NBA) and more mundane ones (e.g., earning a B in math). The key is to ensure that each student identifies a goal that is personally meaningful to them while also being difficult to accomplish.

2.3   Imagining the Ideal Board of Directors

After students have selected appropriate goals, I use a handout to guide them through the creation of their *ideal board of directors* (see Table 1; complete handout included below). Inspired by organizational boards of directors, students are asked to develop a group of people, real or imagined, who can help them achieve their goals. At this point in the exercise, I place no limits on who students can choose for their ideal boards. They may select as many or as few people as they would like to have on their boards, and they can pick people they know personally, people they know about but have never met, or people who may not even exist in real life. Drawing on the discussion about social capital, I ask students to think about the specific knowledge, skills, or resources each board member will provide as they pursue their goals. I remind them of the various forms that social capital can take (material goods, advice, support, etc.), and encourage them to select people who provide a diverse set of resources, rather than people they simply like or would enjoy spending time with. Unburdened by the constraints of reality, students generally have fun with this step in the exercise, selecting celebrities such as Beyoncé to help them learn to sing well enough to earn a part in the school musical, or former President Barack Obama to rally support for a community garden they would like to build.

Table 1 Worksheet used to create the *ideal board of directors*

Board member name	Knowledge/resource/skill

2.4   Creating the Real Board of Directors

The next step in the imaginary board exercise helps students convert their ideal boards into real boards of directors, composed of people who are reasonably likely to help them pursue their goals. Remember that one of the central motivations for developing this exercise was to decrease the cognitive burden of recalling network ties. One of the ways to do this is to narrow the scope of possible responses by offering specific prompts that encourage more accurate recall [10]. Such prompts are embedded in the Knowledge/Resource/Skill column of the ideal board worksheet (Table 2). Without copying the ideal names, I ask students to copy the various resources they sought in their ideal board to a new worksheet (see Table 2). Then, working backwards from the knowledge, resources, and skills they identified as important for achieving their goals, I invite them to identify individuals in their own personal networks who can provide those things. Many students struggle with this conversion at first; however, with a bit of encouragement, they almost always successfully identify appropriate substitutes. For example, a talented cousin might give suitable singing lessons in lieu of Beyoncé, and a city council member might be able to lend support for a community garden if President Obama is not available.

As students work through this portion of the exercise, it is helpful to have them solicit suggestions from the group when they get stuck. Typically, I run this exercise among groups of young people who know one another, and therefore know about one another's networks. Often, students to suggest appropriate board members within one another's networks (e.g., "the school librarian has a list of summer job openings,") or to offer to make introductions to fill holes in peers' boards (e.g., "my cousin works in a café that could hang your art.") It is also not uncommon for me or other event organizers to be reasonable sources of desired knowledge, resources, or skills. Many students would like information about college, for example, something a college professor such as myself can easily provide. I encourage students to leverage these resources, underscoring how we represent weak ties in their networks. (Given the prevalence of requests for follow up conversations about college, I have made a habit of scheduling an "office hour" after presentations to facilitate these interactions.)

Before students finalize their real boards, I ask them to indicate how well they currently know the people they have selected for their boards. Frequently, students will build boards exclusively of people they know very well. I remind them once again of the importance of strong and weak ties and encourage them to diversify their boards,

Table 2 Worksheet used to convert the ideal board to a real board

Board member name	Knowledge/resource/skill	How well do you *currently* know this person?
		Not at all Very well o o o o o
		Not at all Very well o o o o o
		Not at all Very well o o o o o
		Not at all Very well o o o o o

swapping out well-known people for less well-known people if possible. As students complete their real boards of directors, they are often surprised at how many people in their lives can provide the social capital they need to help them achieve their goals.

2.5 Setting an Intention

There is a large body of research that suggests people are more likely to achieve their goals if they create specific intention about when, where, and how they will work on them [19]. To that end, setting an intention may increase the likelihood that students will apply their new knowledge of social networks and social capital after the session has concluded. Therefore, to conclude the exercise, I ask students to select one person from their (real) boards of directors and make a plan for when, where, and how they will contact that person to discuss their goals.

3 Conclusion

The imaginary board of directors exercise introduces students to concepts in social networks and social capital and helps them develop strategies to access social capital in service of their personal, professional, and academic goals. The exercise proceeds in five steps, each of which scaffolds students' existing knowledge to build deeper understandings of essential concepts and how they apply to the students' real lives. The **introduction** teaches students key concepts in social networks and social capital. The **goal** motivates their participation and connects theories of social networks and social capital in students' own lives. The **ideal board** helps students develop a list of desired resources, free from the cognitive burden of recalling their own networks. Using these lists of desired resources to prompt their recall, the **real board** invites students to identify members of their own personal networks who can help them achieve their goals. Finally, the **intention** increases the chances students will put the key lessons from the exercise into practice following the workshop.

Students completing the workshop leave with improved understandings of social networks and social capital, as well as specific plans for how to apply these concepts to improve their chances of achieving personal goals. I have used the imaginary board of directors exercise in a variety of different settings, including as part of a suite of training activities for youth activists and community organizers. Qualitatively, youth report finding the exercise tractable and useful, particularly for helping them reason strategically about the breadth of resources available in their personal social networks. Longitudinal observations by adult program facilitators further suggest that youth do, indeed, follow through on their intentions and doing so helps build youth efficacy around their activism and community organizing goals. Empowering youth to think strategically about the social capital embedded within their social networks may therefore improve their abilities to identify and activate the resources they need to achieve their personal and collective goals.

Appendix: Worksheet

Imaginary Board of Directors

Your Name

Many successful organizations accomplish their goals by creating a board of directors or a small group of people who can support, advise, and inspire them. When they form this group, they are creating a *network* of people and gain access to *social capital* that can help them achieve their goals.

In this exercise, we would like you to create a board of directors for yourself. This exercise has several parts. First, think of a specific goal you would like to accomplish in the future. This should be something challenging that you hope to accomplish for yourself in the next 1–3 years. You can choose an academic goal, an athletic goal, an artistic goal, a career goal, or anything else that matters to you personally.

What Is Your Goal?

Now, we would like you to think about a board of directors who would be useful to YOU PERSONALLY as you work toward your goal. Think about a group of people, real or imagined, living or dead, who can help you achieve your goal. There are no limits on who you can choose for your imaginary board of directors. You may have as many or as few people as you would like on your board. You can pick people you know personally, people you know about but have never met, or people who may not really exist in real life. The key is to pick the best group of people to help you achieve your personal goal.

Ask yourself, what kind of knowledge, resources, and skills will you need to be successful? Who, specifically, could provide those things? Fill in the chart below with ideal members for your board of directors. Remember: "think networks" as you put together your board of directors!

Board member name	Knowledge/resource/skill provided

Now, we would like you to revise your board of directors to include only people you could reasonably meet or interact with. These may be the same people you listed on your ideal board of directors (above), or they may be different people. These do not need to be people you currently know, but they should be people you could possibly come to know.

Sometimes, it will be very easy for you to think of ways to get people to be on your board of directors. Sometimes it may not be obvious why someone would help you out. But it may still be possible to include these people in your personal board of directors. Remember, people tend to do things for people who have something to offer them. Think about what you might be able to offer to the people on your ideal board of directors (above). Think about how you might go about getting each person to join your board of directors. How would you reach out? What might you offer? Why would that person want to help you?

Board member name	Knowledge/resource/skill provided	How well do you CURRENTLY know this person?
		Not at all Very well o o o o o
		Not at all Very well o o o o o
		Not at all Very well o o o o o
		Not at all Very well o o o o o
		Not at all Very well o o o o o
		Not at all Very well o o o o o
		Not at all Very well o o o o o
		Not at all Very well o o o o o

Put a star next to the member of your board of directors who you think will be MOST IMPORTANT for helping you achieve your goal.

Make a plan to reach out to this person – when and where will you contact them and how will you do it?

References

1. DeAndrea, D. C., Ellison, N. B., LaRose, R., Steinfield, C., & Fiore, A. (2012). Serious social media: On the use of social media for improving students' adjustment to college. The Internet and Higher Education, 15(1), 15–23.
2. Marsden, P. V., & Hurlbert, J. S. (1988). Social Resources and Mobility Outcomes: A Replication and Extension. Social Forces, 66(4), 1038–1059. doi:https://doi.org/10.1093/sf/66.4.1038

3. Seibert, S. E., Kraimer, M. L., & Liden, R. C. (2001). A social capital theory of career success. Academy of Management Journal, 44(2), 219–237.
4. Coleman, J. S. (1988). Social capital in the creation of human capital. American Journal of Sociology, S95-S120.
5. Putnam, R. D. (1995). Bowling alone: America's declining social capital. Journal of democracy, 6(1), 65–78.
6. Lin, N. (2002). Social capital: A theory of social structure and action (Vol. 19): Cambridge University Press.
7. Burt, R. S. (2005). Brokerage and Closure: An Introduction to Social Capital. New York, NY: Oxford University Press.
8. Badawy, R. L., Stefanone, M. A., & Brouer, R. L. (2014). I'm Not Just Wasting Time Online! Test of Situational Awareness: An Exploratory Study. Paper presented at the 47th Hawaii International Conference on System Sciences (HICSS) Waikoloa, HI.
9. Krackhardt, D. (1990). Assessing the political landscape: Structure, cognition, and power in organizations. Administrative Science Quarterly, 342–369.
10. Brewer, D. D. (2000). Forgetting in the recall-based elicitation of personal and social networks. Social Networks, 22(1), 29–43. doi:https://doi.org/10.1016/S0378-8733(99)00017-9
11. Hammer, M. (1984). Explorations into the meaning of social network interview data. Social Networks, 6(4), 341–371.
12. Brashears, M. E. (2013). Humans use compression heuristics to improve the recall of social networks. Scientific Reports, 3.
13. Casciaro, T. (1998). Seeing things clearly: Social structure, personality and accuracy in social network perception. Social Networks, 20(4), 331–351.
14. Smith, S. S. (2005). "Don't put my name on it": Social Capital Activation and Job-Finding Assistance among the Black Urban Poor1. American Journal of Sociology, 111(1), 1–57.
15. Mouw, T. (2002). Racial differences in the effects of job contacts: Conflicting evidence from cross-sectional and longitudinal data. Social Science Research, 31(4), 511–538.
16. Smith, E. B., Menon, T., & Thompson, L. (2012). Status differences in the cognitive activation of social networks. Organization Science, 23(1), 67–82.
17. DiMaggio, P., & Garip, F. (2012). Network effects and social inequality. Annual Review of Sociology, 38, 93–118.
18. Uzzi, B., & Dunlap, S. (2005). How to build your network. Harvard Business Review, 83(12), 53.
19. Gollwitzer, P. M., & Sheeran, P. (2006). Implementation intentions and goal achievement: A meta-analysis of effects and processes. Advances in experimental social psychology, 38, 69–119.

Network Visualization Literacy: Novel Approaches to Measurement and Instruction

Angela Zoss, Adam Maltese, Stephen Miles Uzzo, and Katy Börner

1   Introduction

Broadly speaking, we visualize data to investigate complex relationships among variables and to communicate these relationships to others. Network visualizations translate network data into a visual representation of some combination of the actors, relationships, clusters, and data attributes. The value of visualizing the structure of relationships and connections is being recognized as an increasingly important twenty-first-century skill due to the need for all people to have a better understanding of complex phenomena across disciplines.

A call for increased literacy about networks, writ large, resulted in the development of essential concepts that can be taken as a set of goals for what a network-literate person should know by the time they graduate high school [1]. For the purposes of this chapter, we define network visualization literacy (NVL) as the ability to read, interpret, and create visualizations of various types of networks. Research on NVL is still in its early phases, and recent studies suggest that NVL, and more generally data visualization literacy, of youth and adults is not very high or broad [2, 3].

Given this, we focus this chapter on a set of topics that together constitute an attempt to build a more comprehensive vision for NVL, including how to measure NVL, the role of NVL in teaching and learning, what the current research says

A. Zoss (✉)
Duke University, Durham, NC, USA

A. Maltese · K. Börner
Indiana University, Bloomington, IN, USA

S. M. Uzzo
New York Hall of Science, Corona, NY, USA

© Springer International Publishing AG, part of Springer Nature 2018   169
C. B. Cramer et al. (eds.), *Network Science In Education*,
https://doi.org/10.1007/978-3-319-77237-0_11

about NVL across a variety of learning contexts, and recommendations based on our current understanding of best ways to improve NVL. Before we move into discussing these topics, we first define how we conceive of NVL.

2 Network Visualizations

The simplest network visualization is an adjacency list, where each node is itemized and followed by a list of all of the other nodes with which that node shares a link (its neighbors). In the example in Fig. 1a, entity A has connections with B and D, and entity B has connections with A and C. Entities C (with B) and D (with A) only have singular connections with other nodes.

Large networks are more likely to be visualized as matrices or node-link diagrams and can be displayed using one or more of several organizing principles. A matrix visualization (Fig. 1b, representing the same network data as Fig. 1a) is a tabular visualization where a node is represented by either a row or a column (or both) and a link is represented by a numerical value placed in the cell where a node row and a node column intersect. For example, in a matrix visualization of a network of individuals who send text messages to each other, a two-dimensional table is created where the same names appear in the row and column headers. Numerical values representing the number of texts sent between the two people will appear in the cell where the row of one individual and the column of the other intersect. Columns and rows can be ordered to highlight patterns in the data values, such as social cliques where all members text each other a lot [4].

In contrast to a matrix, a node-link diagram represents each actor as a single point using some graphical icon or symbol (often a circle). The presence of a link between two actors is visualized by the addition of a line or arc between the nodes (Fig. 2). These components are often laid out such that smaller distances between nodes represent higher similarity (Fig. 3), but nodes can also be arranged in a circular layout, perhaps in order of a certain property (e.g., a node's number of links), or against a separate reference system like a geospatial map or a science map, where scientific disciplines are arranged in space using citation- or topic-based similarity algorithms (Fig. 4).

Fig. 1 Sample network visualizations: adjacency list (**a**) and matrix (**b**)

A

Node	Neighbors
A	B,D
B	A,C
C	B
D	A

B

	A	B	C	D
A		1	0	1
B	1		1	0
C	0	1		0
D	1	0	0	

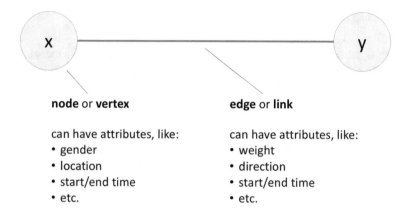

node or vertex   edge or link

can have attributes, like: can have attributes, like:
• gender   • weight
• location   • direction
• start/end time   • start/end time
• etc.   • etc.

Fig. 2 Node-link diagrams typically represent nodes as circles and links as lines or arcs

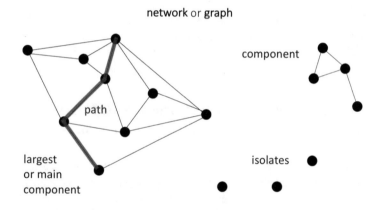

Fig. 3 A simple node-link diagram, labeled with common network-related terminology

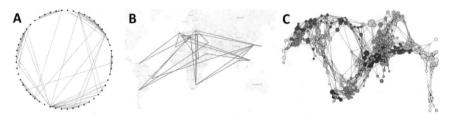

Fig. 4 Sample network visualizations, using a circular layout algorithm (**a**), a geographic layout (**b**), and a science map (**c**)

3 Researching Network Visualization Literacy

In general, we define data visualization literacy as the ability to make meaning from and interpret patterns, trends, and correlations in visual representations of data [5]. In order to interpret visualizations, users need the ability to complete a combination of the following tasks: read text, interpret data arrangements (e.g., to see correlations, trends), and compare object properties (e.g., compare the sizes of nodes in a network given a legend). Users of any information visualization form may engage in a variety of tasks, including both low-level tasks like data foraging and high-level tasks like problem-solving and composing (i.e., making decisions based on data trends) [6].

As a subset of data visualization, network visualization is subject to many of the same kinds of interpretation issues present in other approaches to data visualization. Given the range of abilities needed to interpret visualizations and the myriad tasks possible, it is important to acknowledge the opportunities for network visualizations to be easily misinterpreted. These challenges arise through lack of clarity about the limits of network visualizations in interpreting very complex systems and the many ways that characteristics and behaviors of network components can be represented.

3.1 Representational Literacy: Do Individuals Understand How Network Data Are Converted into Visuals?

There has been an ongoing concern among network science practitioners about the trajectory of network science as a way into deepening data understanding, particularly of large data sets. A call for increased literacy about networks, writ large, has resulted in the articulation of a set of seven essential concepts and core ideas [1]. These essential concepts can be taken as a set of goals for what a network-literate person should know by the time they graduate high school. The fourth essential concept is: "Visualizations can help provide an understanding of networks," and the core ideas subsumed by it include:

- Networks can be visualized in many different ways.
- Diagrams of a network can be drawn by connecting nodes to each other using edges.
- There are a variety of tools available for visualizing networks.
- Visualization of a network often helps to understand it and communicate ideas about connectivity in an intuitive, nontechnical way.
- Creative information design plays a very important role in making an effective visualization.
- It is important to be careful when interpreting and evaluating visualizations because they typically do not tell the whole story about networks.

These essential concepts are relatively new, and scaling them into wide adoption will require robust validation and transformative professional development that integrates new curriculum and learning materials on network visualization into rigorous content knowledge and pedagogical approaches.

Research on the skills required to interpret network visualizations and the prevalence and quality of those skills is still in early phases. Small-scale studies investigated the comprehension of the basic metaphors used by the diagrams [7], the specific structural properties of network data [8], and the graph design aesthetics that are most likely to improve performance on quantitative interpretation tasks [9]. In the sections that follow, we outline some general questions related to NVL that have been investigated through research along with initial findings.

3.2   Metaphoric Literacy: How Intuitive Is the Arrangement of Nodes and Links in a Network Visualization?

Network visualization literacy studies might address whether users understand the metaphoric properties of the visualization, that is, the implicit structures the visualization is using to represent network data. Most node-link diagrams have conventions that guide interpretation of the diagram, such as:

- The positions of nodes are an approximation of the similarity between the nodes, based on an analysis of the links between nodes (and possibly also the weights of those links).
- Nodes that are close together are more similar than nodes that are far apart (the distance-similarity metaphor).
- The positions of nodes may be influenced by aesthetic choices that are encoded into the layout algorithm (e.g., to minimize edge crossing) or that are used to make manual adjustments (e.g., to eliminate overlap of two nodes by manual shifting).
- Network visualizations can be rotated or reflected in space arbitrarily.
- Some network visualizations omit a portion of the links to better focus attention on the node positions and the most important link structures.
- Shorter links are usually stronger than longer links, even though longer links may draw the eye and shorter links may be so short that they almost disappear.

Studies of the metaphoric properties of network visualizations are rare and have focused primarily on the distance-similar metaphor. Fabrikant and colleagues [7, 10, 11] explored the judgments of novice users of network visualizations regarding the presumed similarity of two pairs of target nodes, manipulating a variety of topological and aesthetic variables: the Euclidean distance between the nodes, the cumulative measured length of links between the nodes, the number of intervening nodes on the path between the target nodes, and the width, darkness or hue of links. In all studies, participants overwhelmingly associated similarity with the length of the path between two nodes (in terms of geometric length or direct-line distance, not the number of links in the path). Nodes close to each other "as the crow flies" were considered less similar to each other than nodes that had a shorter network connection. The only design features of a network that contradicted this powerful intuition were the width of a link and, to a lesser extent, the darkness of a link; wider links especially made nodes seem more similar to each other, even if those nodes had a longer measured path.

In a final study, [10] compared judgments of node similarity to judgments of node distance by making a slight change to the task instructions from their previous studies, such that participants answered questions about similarity and distance separately. When asked about distance, participants focused on Euclidean distance. When asked about similarity, participants focused on the geometric length of links. These results are encouraging, in that network layout algorithms may make compromises about where a node is positioned, thereby rendering "as the crow flies" distances less meaningful than the presence of links. On the other hand, the length of links can be determined both by the layout of the nodes and by whether the layout algorithm has a constraint on link length. Novices without a sophisticated understanding of layout algorithms will be likely to make judgments based on the length of the lines.

One way of interpreting these findings is through the lens of basic perceptual skills. Even without special training, users of visualizations have natural skills for interpreting spatial information. These skills were described over a century ago by German psychologists as "Gestalt laws" [12], and they can help explain how components of data visualizations are understood on a very fundamental, perceptual level. These laws are especially relevant for network visualizations, where training and even exposure are uncommon among a general population. The arrangements of nodes in space and the connection of those nodes by lines have very strong connotations for users, and visualization designers must anticipate how that will affect interpretation.

3.3 Topological Literacy: Do Individuals Understand Basic Network Properties When Reading Network Visualizations?

Beyond a user's intuition about a network visualization, researchers may also want to investigate whether users can glean topological information (i.e., the mathematical or statistical properties of the network data underlying the diagram) from the visualization. Depending on the field of study, different parts of a network dataset may be considered especially important. In some fields, the clusters of nodes in the network are most important, whereas in other fields it is important to identify specific nodes that are highly influential. For example, a study could measure a user's ability to identify nodes with a high betweenness centrality score (i.e., those nodes that lie on heavily traveled paths between node clusters) from the network visualization. Testing whether a user can read or estimate topological information about network data from a visualization can be an important way of assessing either the user's literacy or the visualization's success.

A user's topological literacy is dependent on many factors: the user's prior training with both network data and network visualizations, the choice of network layout algorithm (and by extension, the topological properties that are emphasized by the network layout algorithm), any additional design choices made by the producer of the network visualization (e.g., adding color coding to emphasize a particular topological property), the specific properties of that particular network dataset (e.g., the size of the network, whether some nodes have notably more links than others), and the choice of topological property (i.e., "task") to measure.

3.3.1 Effect of Layout or Base Map Choice

Node-link diagrams have a wide variety of layout algorithms (Fig. 5) that determine the position of nodes and edges. The most common layout algorithms, especially for small- or medium-sized networks, are algorithms drawn from physical analogies like springs and forces, pushing and pulling the nodes into place based on the presence and/or weight of edges. The complexity of network data means that there is no one "correct" layout of the nodes and edges – two nodes may have a strong link to each other, but they may also be strongly connected to other nodes that are very far apart. Because of this complexity problem, different layout algorithms have been developed either to prioritize different features of the data or to make certain types of visual judgments easier.

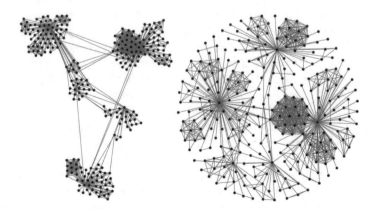

Fig. 5 Two visualizations of the same network data. The layout algorithm on the left prioritizes clusters, while the layout algorithm on the right prioritizes even node distribution

One way of evaluating layout algorithms is to explore the extent to which the layout follows guidelines for graph design aesthetics [13, 14]. Many such graph aesthetic principles have been identified [9], including:

- Global and local symmetry
- Non-overlapping nodes
- Minimized edge crossings
- Edges of equal length
- Evenly spaced nodes
- Visual representation or emphasis of clusters (e.g., intra-cluster edges are short-ened; inter-cluster edges are lengthened)
- Space-filling algorithms
- Node-area awareness

The most widely studied aesthetic properties for network visualizations have been edge crossings and the angles created by those crossings. These features have been found to have a large impact on topological literacy. Different layout algorithms also vary greatly in their performance on these aesthetic properties.

Seminal work by Purchase and colleagues [15–19] manipulated and tested a series of aesthetic properties of network visualizations to determine user performance on three tasks for finding: (a) the length of the shortest path between two nodes, (b) the number of nodes that need to be removed to destroy a path between two nodes, and (c) the number of edges that need to be removed to destroy a path between two nodes. Through a sequence of related studies, Purchase and colleagues [15–18] systematically investigated the effects of edge bends, edge crossings, layout symmetry, angles between links as they leave a node, and use of an orthogonal grid for nodes and links. Results consistently emphasized that higher numbers of edge crossings and high numbers of edge bends generally reduce performance, measured via task accuracy and response time. Related work by Huang and colleagues [20–24] supports these results and suggests that edge crossings with small angles, especially, inhibit performance (measured by accuracy, response time, and self-reported mental effort) on tasks that require users to follow paths.

As a follow-up to the original studies by Purchase and colleagues, [19] introduced the concept of path continuity or the lack of abrupt changes in direction of the path. This study focused on a single task – length of the shortest path between two nodes – and found that response time on this task increased as a result of the following changes (in order of influence): increase in number of edges in shortest path, decrease in continuity of shortest path, increase in number of crossings on shortest path, and increase in number of branches off nodes in the shortest path. This suggests that for tasks requiring users to follow a path, anything increasing the number of additional candidate paths or that makes it harder to focus on the shortest path will increase the time needed to complete the task. Rather than focusing on properties of the entire network visualization, it may make more sense to optimize visualizations for specific tasks and the aesthetic properties that will make those tasks easier.

3.3.2   Effect of Data Overlay Design Choices

Regardless of the layout algorithm used, other basic graphic design properties that apply to all visualizations should be considered when designing network visualizations. A series of core perceptual studies have addressed basic human perceptual abilities as they relate to interpreting information visualizations. Both early studies and more recent replications [25, 26] suggest that humans have aptitude for comparisons related to position in space and length of an object. Accuracy suffers when tasks require comparisons of area (like comparisons between two circles) or color value (like the comparisons between two shades of red). Node-link diagrams employ only relative positioning, and even those relative positions are the result of algorithms that may not have an optimal solution for a 2D visualization. In data visualization, there is often a tension between accuracy and aesthetics. The differences in positions in a node-link diagram are not meant to be interpreted with great accuracy, despite human acuity for position comparisons. Conversely, node-link diagrams often employ size and color coding to emphasize topological data in the visualization despite our relatively low acuity with those visual encodings. These mismatches between node-link diagrams and our basic human perceptual systems suggest challenges for the use of network visualizations without supplemental numerical information.

3.3.3   Effect of Network Data Properties

The basic properties of a network dataset can also have a large impact on the effectiveness of the visualization. Ghoniem, Fekete, and Castagliola [8] compared task performance of users viewing matrices and node-link diagrams, varying the size and densities of sample data sets. They found performance on all experimental tasks deteriorated for node-link diagrams as the size increased from 20 nodes to 50 nodes and again between 50 nodes and 100 nodes. Increases in density between 0.2 and 0.6 had mixed effects on task performance. They concluded that certain tasks are much harder with high-density networks, while others show no significant drop in accuracy as density increases. Similarly, Purchase, Cohen, and James [18] found that an increase in density of node-link diagrams relates to a decrease in accuracy on tasks dealing with the connectivity of a network.

4   Teaching Network Visualization Literacy

As reviewed above, research within the visualization community focuses primarily on experimental studies of network visualization comprehension, limited to specific predetermined tasks. A more robust understanding of network visualization literacy must also take into account both how users understand network visualizations when they encounter them in their daily life and how individuals can gain the expertise necessary to produce their own network visualizations. Thus, a combination of formal and informal education is desirable for empowering many to read and make network data visualizations. Here we present and discuss three existing approaches–Connections: The Nature of Networks (a public science museum exhibition at the New York Hall of Science), NetSci High (a research program for high school students sponsored by Boston University, Binghamton University, USMA West Point, and the New York Hall of Science), and the Information Visualization MOOC course at Indiana University.

4.1 Network Visualization in Informal Learning Environments

Informal learning environments (which includes unstructured learning opportunities such as museums and personal learning) provide opportunities for acquainting the public with network visualization to increase NVL. Because of the unstructured nature of these environments and the relative novelty of the use of network visualization as a tool for understanding complex systems, significant scaffolding is required for effective learning and knowledge transfer.

4.1.1 Connections: The Nature of Networks

The first public museum exhibition on network science was developed by the New York Hall of Science in 2004 [27]. The pedagogical goal of this exhibition was to acquaint museum visitors with the fundamentals of network science, including the basic ways networks are represented as a series of links and nodes, as well as the generalizability and value of how most kinds of complex connected systems can be represented as networks, the benefits of these kinds of representations, and a basic characterization of complex network concepts (small worlds, scale-free properties and emergence). To address a very diverse audience (including all ages), it was theorized that the experience overall should engage visitors in network concepts in a variety of ways, including visual representations, sound, and embodied or physical interaction with networks and network concepts (Fig. 6).

A significant challenge to developing this experience was that the notion of networks as a general principle was a new idea to visitors. In a preliminary visitor study [28], visitors could readily identify computer and communications networks with little or no prompting, but they could not readily identify networks in either social or natural contexts. A summative evaluation of the exhibition [29] indicated an increase in the number of visitors who identified networks in a broad spectrum of applications, particularly environmental and social, and indicated that networks are a way to understand the world through the Connections exhibition. To achieve these outcomes, however, required intervention by floor staff to explain relevant network ideas represented in the exhibition. By far the most popular and effective aspects of the experience were when visitors physically interacted with networks.

Fig. 6 *Ropes and Pulleys* (left) convey the complexity and dynamic of networks. Visitors turn the wheels to change the topology of the pulleys and ropes creating clusters and isolated nodes. NEAR (bottom right) simulates the dynamics of social networks using nearest neighbor algorithms

On balance, beyond positive affect and recognition of the ubiquity of networks, there was little transfer of knowledge. It was a much more difficult task for visitors to deepen their understanding of the properties of networks. The implications of this work for NVL are that effective engagement of museum visitors in complex network ideas (a) is deeper when visualization is combined with hands-on activities in which the visitor is engaged in network concepts, (b) requires intervention of floor staff for understanding specific network properties, and (c) exists within a reality of an overall lack of understanding of networks among a diverse visitorship, indicating the need for more learning opportunities for the public about the importance and utility of networks.

4.2   Network Visualization in Formal Learning Environments

For the sake of this chapter, formal learning environments circumscribe teaching and learning in primary, secondary, and post-secondary school-based environments. The following section details NVL activities in both university and high school classrooms.

4.2.1 NetSci High

Network visualizations have been taught at the university level for some time, but secondary educational environments only recently started exploring this topic. As with the introduction of any new idea, finding a way to fit networks into existing curriculum is difficult for teachers, who are accountable primarily for their student's performance on standardized tests. Because networks align well with mathematics and science content standards through the Common Core [30] and NGSS [31], it is reasonable to infuse these ideas into curriculum, create professional development opportunities, and also student and teacher research through guidance by researchers and university faculty.

NetSci High started in 2010 and is a first of its kind program to train high school teachers and students in network analysis techniques and have them apply it to research mentored by university researchers and graduate students [32]. An important aspect of this program is the training of students and teachers, which between 2012 and 2015 took the form of a 2-week "boot camp" at Boston University, in which teachers and students were immersed in network concepts and trained in tools to equip them for mentored research during the school year. Research projects culminate in display and defense of research at the International School and Conference for Network Science.

Findings from the evaluation of the Connections exhibition at the New York Hall of Science were useful to inform the development of the training and the role of network visualization in the program. Specifically, this involved exposing participants to a wide variety of applications of networks and how they are visualized, as well as providing hands-on and embodied ways to demonstrate and engage with network concepts. Central to the training was skills development in the use of network analysis tools. NetworkX and Gephi were the primary tools taught, and various other analysis environments and techniques were included in the actual research phase. Program evaluation by Davis Square Research Associates [33] indicated significant gains in the understanding of (a) the value of network visualization and its role in analysis of complex networks; (b) the intimate relationship between analysis and visualization; (c) the process of representing a variety of network attributes, which can be accomplished through a variety of tools; and (d) the importance of an intensive approach to teaching novices network visualization as a tool to analyze and communicate findings in network science.

4.2.2 Information Visualization MOOC

In a long-running teaching and research program at Indiana University, teaching university students to understand and create network visualizations began by developing a systematic process for designing effective visualizations. This framework for creating visualizations has then been embedded into course structure, books, activities, software, and digital teaching aids, all of which allowed the graduate-level Information Visualization course to expand into a massive open online course (MOOC).

Frameworks for Network Visualization Education

Börner [34] proposes a general process for converting data into a visualization (Fig. 7), each step of which is based on an analysis of what the users of the visualization need and want from the visualization. First, data need to be parsed and read (READ).

Extensive cleaning and preprocessing might be needed. Temporal, geospatial, topical, and network analyses might be performed to identify trends and patterns (ANALYZE). The visualization phase (VISUALIZE) comprises three major steps. First, the appropriate reference system must be identified. This reference system becomes the stable base map onto which data are layered. Second, the reference system might be modified (e.g., an axis may undergo a logarithmic transformation). Third, additional data variables are visually encoded using diverse graphic variable types. Ultimately, the visualization must be deployed (DEPLOY) (i.e., printed, published online, etc.) Last but not least, the visualization is presented to stakeholders for validation and interpretation. Frequently, new visual insights lead to new questions, requiring additional data analysis and visualization – the cycle repeats.

This detailed formulation of different steps involved in visualization design open each step up for the critical discussions that are necessary for gaining data visualization literacy. For example, different needs from stakeholders, combined with different properties of the data, will lead to different visual design recommendations. In the case of network visualizations, the framework is especially helpful in guiding students through a series of design choices that are easy to dismiss as arbitrary because of the lack of standardized guidelines and training within the broader information visualization community.

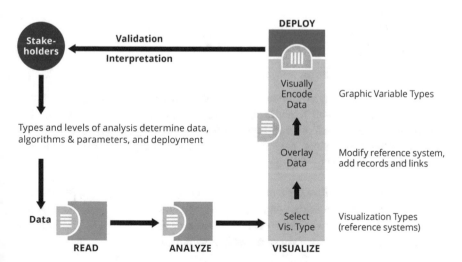

Fig. 7 Needs-driven workflow design with science map network example on right

The information visualization MOOC (IVMOOC) [35] is a graduate-level course that has been continually developed and taught at Indiana University (IU) since 2013. Most students, coming from over 100 countries, take the course for free to earn a personalized letter of accomplishment and digital Mozilla badge. Additionally, in Spring 2016, more than 120 students registered for three IU credits as part of the Information and Library Science M.S. program and the online Data Science M.S. program offered by the School of Informatics and Computing.

Course Structure

The IVMOOC course aims to improve data visualization literacy – the expertise and skills needed to read and make data visualizations. It teaches theoretical foundations and advanced tools that help turn data into insights.

The course uses a combination of hands-on case studies showing how to read, analyze, and visualize; theory lectures; client projects; homework assignments; and exams to empower students to design effective visualizations that take the needs of users into account. In the first week of the course, students are introduced to the visualization framework, which is used to structure the course's schedule and exams, textbook [36], tools, and an IVMOOC flashcard app (discussed more below). In weeks two to six students use the framework to learn about a variety of types of visualizations, including network visualizations. In the last 7 weeks of the course, students collaborate on real-world projects for a variety of clients. Results from previous student projects are published in [36].

Each unit includes theory and hands-on sections. Each theory section comprises:

- Examples of exemplar visualizations
- Visualization goals
- Key terminology
- General visualization types and their names
- Workflow design
- Discussion of specific algorithms

Each hands-on section guides students through user and task analysis; data preparation, analysis, and visualization; deployment; and the interpretation of visualizations. The sections feature in-depth instruction on how to navigate and operate several software programs used to visualize information. Furthermore, students learn the skills needed to visualize their own data, allowing them to create unique visualizations.

The theory and hands-on components are standalone, meaning that participants can read/watch whichever section they are more interested in first, and then review the other section. After the theory videos, there are self-assessments, and after the hands-on videos, there are short homework assignments.

Textbook

The *Visual Insights* textbook [36] was designed as a companion resource for students taking the IVMOOC. It contains all theory and workflows covered in the course. While the *Atlas of Knowledge* [34] aims to feature timeless knowledge, or principles that are indifferent to culture, gender, nationality, or history, the IVMOOC and associated textbook cover "timely knowledge," or the most current data formats, tools, and workflows used to convert data into insights.

Analogous to the IVMOOC course, Chap. 1 introduces the visualization framework intended to help non-experts assemble advanced analysis workflows and design different visualization layers. It also showcases how the framework can be applied to "dissect visualizations" for optimization or interpretation. Chapters 2–7 in the textbook introduce the different types of analysis: temporal (when), geospatial (where), topical (what), and trees and networks (with whom). Chapter 8 presents exemplary case studies that resulted from IVMOOC real-world client projects.

Software

Every student who registers for the IVMOOC gets experience using the Sci2 Tool [37] a software application for data analysis and visualization developed by Börner at IU. The NSF-funded tool has been in development since 2008 and benefits from more than 10 years of tool development and feedback from many of the more than 150,000 tool users in academia, industry, and government. The tool supports the temporal, geospatial, topical, and network analysis and visualization of scholarly datasets at the micro (individual), meso (local), and macro (global) levels. It implements the visualization framework to help users assemble more than 180 algorithms into proper workflows. Specifically, it organizes the main menu structure by workflow steps (from reading and preprocessing data to analyzing and visualizing data and saving out results) and by visual analysis type (temporal, geospatial, topical, networks) using the visualization framework discussed above.

Flashcard App

Visualization designers and users must have a basic understanding of different visualizations – their types and the visual encodings used. They must be able to recognize and name visualizations in order to refer to and talk about visualizations. The IVMOOC Flashcard app lets users browse more than 60 information visualizations. Users swipe to navigate through visualizations, pinch in/out to zoom, and tap to turn the card to access information about name of the visualization, visualization type (e.g., graph, map, network layout), visual encoding used (graphic symbol types and graphic variable types), and reference to additional information provided in the *Atlas of Knowledge* [34]. The Flashcard app, created in Unity 3D, supports both Android and iOS.

5 What Are our Best Ways Forward?

Based on our review of relevant research and experiences with teaching and learning with network visualization in formal and informal settings, we make the following recommendations for improving network visualization literacy:

- Use data that are meaningful to the learner. Data that are personally relevant to, directly collected from/by or selected by the learner will increase their engagement and familiarity with the data. Similarly, when asking questions of the data, instructors are urged to pick tasks that make sense for network visualizations but also those that make sense for the selected data and for the research questions the user is investigating.
- When introducing novices to networks and visualization techniques, best practices would suggest the use of small networks with low density to increase understanding.
- When designing visualizations, leverage core perception mechanisms by following Gestalt grouping, continuity, and proximity principles. Additionally, endorsing certain types of aesthetic principles like minimal edge crossing and path continuity should improve the likelihood of understanding by the public.
- Following best instructional practices, educators should engage novices in low complexity tasks with greater support and move toward higher-complexity tasks and networks, withdrawing support as the learners gain competence.
- Network scientists need to provide explicit instruction on how readers/users should read the visualizations they create. Clearly outlining what conclusions are and are not valid will help users in interpretation. This should include use of standardized terminologies to leverage prior knowledge.

6 Discussion

In conclusion, there is ample evidence that network visualization is an important tool for understanding complex connected systems, but it is important that it be thoughtfully combined with other pathways into understanding what networks are, their characteristics and behaviors.

To enact these recommendations and advance the nascent research on NVL, we invite close collaboration with others on developing both widely adoptable visualization frameworks that can be used to teach information visualization theory and methods and also custom development of a more refined and meaningful definition and framework for NVL. Efforts must be made to develop guidelines that recommend skills and learning outcomes and competencies for both learners that have taken information visualization courses in formal settings and a wide audience of citizens and policymakers. Additionally, the visualization community should work together openly to standardize terminology, theoretical frameworks, and visualization techniques. This work should involve the development, testing, and implementation of course designs, tools, materials, and activities to increase student competency with interpreting and implementing visualizations, preparing them to evangelize these methods and practices in research, practice, and training. Finally, open data, open code, and open education are true enablers that can empower anyone to convert data into visual insights.

Acknowledgments This work was partially supported by the National Institutes of Health under awards P01 AG039347 and U01CA198934 and the National Science Foundation under awards NCSE 1538763, EAGER 1566393, NRT 1735095, AISL 1713567, and NCN CP Supplement 1553044. Any opinions, findings, and conclusions or recommendations expressed in this material are those of the author(s) and do not necessarily reflect the views of the National Science Foundation.

References

1. Sayama, H., Cramer, C., Porter, M. A., Sheetz, L., & Uzzo, S. (2016). What are essential concepts about networks? Journal of Complex Networks, 4(3), 457–474. doi:https://doi.org/10.1093/comnet/cnv028
2. Börner, K., Balliet, R., Maltese, A. V., Uzzo, S. M., & Heimlich, J. E. (2015). Meaning Making Through Data Representation Construction and Deconstruction. Paper presented at the AERA 2015 Annual Meeting, Chicago, IL.
3. Maltese, A. V., Harsh, J. A., & Svetina, D. (2015). Data Visualization Literacy: Investigating Data Interpretation Along the Novice—Expert Continuum. Journal of College Science Teaching, 45(1), 84–90.
4. Eliassi-Rad, T., & Henderson, K. (2010). Literature search through mixed-membership community discovery. In S.-K. Chai, J. Salerno, & P. L. Mabry (Eds.), Advances in Social Computing: Third International Conference on Social Computing, Behavioral Modeling and Prediction, SBP10 (pp. 70–78). Bethesda, MD: Springer.
5. Börner, K., Maltese, A. V., Balliet, R., & Heimlich, J. E. (2016). Investigating Aspects of Data Visualization Literacy Using 20 Information Visualizations and 273 Science Museum Visitors. Information Visualization, 15(3), 198–213.
6. Card, S., Mackinlay, J. D., & Shneiderman, B. (1999). Readings in information visualization : using vision to think. San Francisco: Morgan Kaufmann Publishers.
7. Fabrikant, S. I., Montello, D. R., Ruocco, M., & Middleton, R. S. (2004). The Distance-Similarity Metaphor in Network-Display Spatializations. Cartography and Geographic Information Science, 31(4), 237–252.
8. Ghoniem, M., Fekete, J.-D., & Castagliola, P. (2005). On the readability of graphs using node-link and matrix-based representations: a controlled experiment and statistical analysis. Information Visualization, 4(2), 114–135. doi:https://doi.org/10.1057/palgrave.ivs.9500092
9. Bennett, C., Ryall, J., Spalteholz, L., & Gooch, A. (2007). The Aesthetics of Graph Visualization. In D. W. Cunningham, G. Meyer, & L. Neumann (Eds.), Computational Aesthetics in Graphics, Visualization, and Imaging (pp. 57–64): The Eurographics Association.
10. Fabrikant, S. I., & Montello, D. R. (2008). The effect of instructions on distance and similarity judgements in information spatializations. International Journal of Geographical Information Science, 22(4), 463–478. doi:https://doi.org/10.1080/13658810701517096
11. Fabrikant, S. I., Ruocco, M., Middleton, R., Montello, D. R., & Jörgensen, C. (2002). The first law of cognitive geography: Distance and similarity in semantic space. Paper presented at the GIScience 2002, Boulder, CO.

12. Ware, C. (2013). Information visualization: perception for design (3rd ed.). Waltham, MA: Morgan Kaufmann Publishers.
13. Börner, K., Chen, C., & Boyack, K. W. (2003). Visualizing knowledge domains. Annual Review of Information Science and Technology, 37(1), 179–255.
14. Brandes, U. (2001). A Faster Algorithm for Betweenness Centrality. Journal of Mathematical Sociology, 25, 163–177.
15. Purchase, H. C. (1997). Which aesthetic has the greatest effect on human understanding? Paper presented at the Graph Drawing. GD 1997. Lecture Notes in Computer Science, vol 1353.
16. Purchase, H. C. (2000). Effective information visualisation: a study of graph drawing aesthetics and algorithms. Interacting with Computers, 13(2), 147–162.
17. Purchase, H. C., Carrington, D., & Allder, J. (2002). Empirical evaluation of aesthetics-based graph layout. Empirical Software Engineering, 7(3), 233–255.
18. Purchase, H. C., Cohen, R. F., & James, M. I. (1997). An experimental study of the basis for graph drawing algorithms. Journal of Experimental Algorithmics, 2, No. 4. doi:https://doi.org/10.1145/264216.264222
19. Ware, C., Purchase, H. C., Colpoys, L., & McGill, M. (2002). Cognitive measurements of graph aesthetics. Information Visualization, 1(2), 103–110. doi:https://doi.org/10.1057/palgrave.ivs.9500013
20. Huang, W. (2013). An aggregation-based overall quality measurement for visualization. Retrieved from https://arxiv.org/abs/1306.2404
21. Huang, W. (2014). Evaluating overall quality of graph visualizations indirectly and directly. In W. Huang (Ed.), Handbook of Human Centric Visualization (pp. 373–390). New York: Springer-Verlag.
22. Huang, W., Eades, P., Hong, S.-H., & Lin, C.-C. (2013). Improving multiple aesthetics produces better graph drawings. Journal of Visual Languages & Computing, 24(4), 262–272. doi:https://doi.org/10.1016/j.jvlc.2011.12.002
23. Huang, W., & Huang, M. L. (2011). Exploring the relative importance of number of edge crossings and size of crossing angles: A quantitative perspective. International Journal of Advanced Intelligence, 3(1), 25–42.
24. Huang, W., Huang, M. L., & Lin, C.-C. (2016). Evaluating overall quality of graph visualizations based on aesthetics aggregation. Information Sciences, 330, 444–454.
25. Cleveland, W. S., & McGill, R. (1985). Graphical perception and graphical methods for analyzing scientific data. Science, 299(4716), 828–833.
26. Heer, J., & Bostock, M. (2010). Crowdsourcing graphical perception: using mechanical turk to assess visualization design. Paper presented at the Proceedings of the SIGCHI Conference on Human Factors in Computing Systems, New York, NY, USA. http://dl.acm.org/citation.cfm?doid=1753326.1753357
27. Uzzo, S., & Siegel, E. (2010). Connections: The Nature of Networks, Communicating Complex and Emerging Science. In A. Filippoupoliti (Ed.), Science Exhibitions, Communication and Evaluation. Edinburgh: MuseumsEtc.
28. Cohen, S. (2002). Connections The Nature of Networks: Front End Evaluation. Retrieved from Program Evaluation and Research Group:
29. Rothenberg, M., & Hart, J. (2006). Analysis of Visitor Experience in the Exhibition Connections: the Nature of Networks at the New York Hall of Science. Retrieved from Northampton, MA:
30. National Governors Association Center for Best Practices, & Council of Chief State School Officers. (2010). Common Core State Standards. Retrieved from http://www.corestandards.org/
31. NGSS Lead States. (2013). Next Generation Science Standards: For States, By States. Washington, DC: The National Academies Press.
32. Cramer, C., Sheetz, L., Sayama, H., Trunfio, P., Stanley, H. E., & Uzzo, S. (2015). NetSci High: Bringing Network Science Research to High Schools. Paper presented at the Complex Networks VI: Proceedings of the 6th Workshop on Complex Networks CompleNet 2015, New York.
33. Faux, R. (2015). Evaluation of the NetSci High ITEST Project: Summative Report. Retrieved from Boston:

34. Börner, K. (2015). Atlas of knowledge: Anyone can map. Cambridge, MA: MIT Press.
35. CNS Center at Indiana University. (2017). IVMOOC: Information Visualization MOOC 2017. Retrieved from http://ivmooc.cns.iu.edu
36. Börner, K., & Polley, D. E. (2014). Visual Insights: A Practical Guide to Making Sense of Data. Cambridge, MA: The MIT Press.
37. Sci2 Team. (2009). Science of Science (Sci2) Tool. Indiana University and SciTech Strategies, http://sci2.cns.iu.edu.

Network Science in Your Pocket

Toshihiro Tanizawa

1 Introduction

Network science has its root in mathematical graph theory, which began from the famous problem "Seven Bridges of Königsberg," solved by a mathematician, Leonhard Euler, in 1736. Erdös and Rényi combined classical graph theory with modern probability theory and initiated random graph theory in 1959 [1]. Starting in 1998, it became evident that mathematical graph theory could be applied for understanding real-world phenomena, to develop into "network science" in the present context [2, 3]. Network science has now become one of the most active research fields, and a vast amount of knowledge has been accumulated.

Any system that can be mapped onto a structure made from nodes and links is considered to be a network. This ultimate simplicity is the most valuable asset for the concept of "networks" since, in this sense, networks are everywhere. By representing a complicated system as a network, we can analyze the problems relating to the system theoretically and/or numerically. The results are compared to the observed data generated by the system, and the validity of the network representation of the system is verified quantitatively. The methods in network science have thus become powerful tools for understanding various phenomena produced by complex systems.

One of the remarkable features of network science is that, though the cutting-edge results are obtained by highly mathematical and statistical analysis, the most basic concepts, such as nodes and links, degree of nodes, average path length, clustering, and small-world properties, are conceivable even for elementary school children through vivid network visualization without any mathematical expressions. Thus, we believe that the basic knowledge of network science can be included in

T. Tanizawa (✉)
National Institute of Technology, Kochi College, Kōchi, Japan

© Springer International Publishing AG, part of Springer Nature 2018

189

C. B. Cramer et al. (eds.), *Network Science In Education*,
https://doi.org/10.1007/978-3-319-77237-0_12

core curricula at every educational level, from preschool to undergraduate. The concept of "network literacy"[1] embodies this situation [4].

Since most of the experts in network science at present are confined in research universities, institutes, or corporation laboratories, increasing the number of people familiar with the concepts of network science outside academia is the first important step toward the goal of network literacy. In this regard, outreach education activities for K-12 students and/or their teachers led by network scientists are crucial for spreading the concept of network literacy. Moreover, it is preferable that the activities be performed at the places where the teachers are engaged in their everyday teaching, since, by doing so, teachers could realize that network science is a subject that can be taught outside research settings.

However, preparing for outreach activities outside campus can be tedious and troublesome. In the first place, instructors have to prepare various kinds of lecture materials, such as viewgraphs, images, and charts. Since the networks dealt with in network science are dynamic objects on which various phenomena occur, the presentation of the contents should also be dynamic and, preferably, interactive. If the attendees have sufficient skills for software programming, it would be a good option that the lectures include a short hands-on course for developing small applications of network science using various programming languages, such as C, C++, Java, Python, and so on. The instructors, however, would not be able to expect such programming environments at off-campus activities. Connection to the Internet is also preferable, since students may look up various technical terms online during the lectures. The instructors may also use online contents during the lectures. To deal with all of these problems coherently would be a burden for the instructors in preparation of the lectures.

To make this easier, it would be very helpful if instructors have the necessary tools and materials on a stand-alone environment in one tiny box that can be brought anywhere easily, which has become possible through recent advancements in tiny single-board computer technology. The project, "Network Science in Your Pocket" (the Pocket project, hereafter), intends to realize such a "tool box" that enables us to practice outreach activities for network literacy in everyday teaching in the classroom.

This chapter is organized as follows. In the next section, we clarify the requirements for such a toolbox, which is called the Pocket Server in this chapter. Section 3 describes two operation modes of the Pocket Server. In Sect. 4, concrete implementation processes for building the Pocket Server are described. Section 5 reports the details of a lecture for network science using the Pocket Server, held in a technical college in Japan. Section 6 summarized this chapter.

---

[1] A PDF of Network Literacy: Essential Concepts and Core Ideas, in the original English as well as over 20 additional translations, can be downloaded from: https://sites.google.com/a/binghamton.edu/netscied/teaching-learning/network-concepts/.

2 Requirements for the "Tool Box"

In Fig. 1, we show the education levels, which the Pocket project intends to cover. In preparation for a series of lectures in an outreach activity, instructors have to take into account prior knowledge and skills of the students. For instance, instructors would not be able to expect primary school students to have sufficient skill in fluent keyboard typing. The contents of the lectures would thus be preferable if they are presented with a lot of graphics on-screen and proceed by mouse clicking and/or dragging. The most convenient interface for such a style will be a web browser, so we refer to this style as the "browser-oriented" style. As the education level becomes higher, the typing skill of the students improves. Instructors are even allowed to expect that the students might have some knowledge or experiences in software programming. In this case, the contents of the lectures are allowed to contain more materials that require inputs from the students through a character-based user interface, such as a terminal window provided in almost all operation systems presently available. We refer to this style as the "command-line oriented" style. The toolbox of the Pocket project is thus required to be able to deal with both browser-oriented and command-line oriented styles.

To make lectures interactive, the toolbox should accept various types of communication channels and protocols over the local area network (LAN) and the Internet alike. For effective storage of a large amount of data used in the course, a suitable database application might also be required. If a set of lectures for high school or undergraduate students contains assignments of writing progress reports, it might be preferable that the tool box provides a suitable document processing suite such as LaTeX that can handle mathematical expressions, reference citations, indexing, and so on, easily and coherently. For efficient code writing in programming, good old editors such as Vim or Emacs should not to be omitted from a fundamental set of requirements for the toolbox.

Figure 2 is a schematic picture of the toolbox of the Pocket project, the "Pocket Server" that fulfills all of these requirements. In this picture, the Pocket Server is intended to work in two different operation modes alike; one acts as a server that

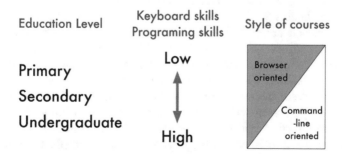

Fig. 1 Education levels from primary to undergraduate over which this Pocket project intends to cover

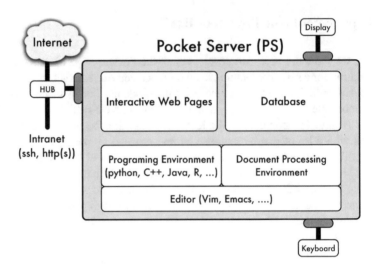

Fig. 2 A schematic picture of a sample structure of the "Pocket Server" (PS)

accepts connection requests from outside and delivers the contents of the course, and the other acts as a client computer for each student in which all necessary tools for studying network science are included in a box.

3   Two Operation Modes of the Pocket Server

As mentioned above, the Pocket Server has two operation modes, the server mode and the client mode. The server mode is the basic mode of the Pocket Server, in which a single Pocket Server works as an ordinary http server. This mode is suitable for giving a browser-oriented course in a lecture room that already provides a set of personal computers for students with a working local area network (LAN). These days, many primary education facilities as well as those for secondary or higher education are equipped with such lecture rooms. In this case, the instructor needs to only bring the Pocket Server and connect it to the screen and LAN of the lecture room. The students access and retrieve the contents of the lectures in the Pocket Server through http connection over LAN. If the server accepts wireless connections, personal computers are not necessary. The attendees can access the contents with their own laptops, tablets, or smartphones. In Fig. 3, we show a schematic picture of the use of the server mode. In case the number of the computers that connect to a single Pocket Server is large and exceeds the capability for handling all requests from the client machines in the lecture room, increasing the number of Pocket Servers for distributed request processing could be an effective treatment.

If the organizer of the lectures can afford to provide a sufficient number of Pocket Servers, each Pocket Server can be operated in the client mode as a stand-alone PC

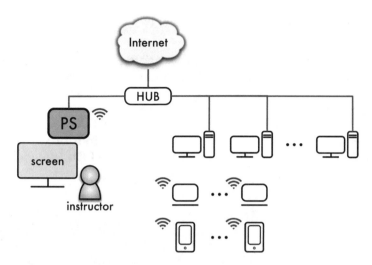

Fig. 3 The use of the server mode

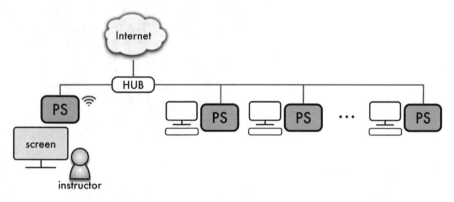

Fig. 4 The use of the client mode

by connecting it to a monitor and a keyboard. (See Fig. 4.) In this case, each student has the same environment as that of the instructor at the start of the session. The students are first encouraged to try the examples presented by the instructor, typing the inputs with their own hands, and see the results. By modifying these examples, the students can start to explore the various topics in network science by themselves. The client mode is thus suitable for command-line oriented lectures. It should be noted, however, that even when most of the contents of the course takes the command-line oriented style, it is always possible for each client to connect to the Pocket Server of the instructor separately. The lectures can thus shift back and forth freely between the server mode and the client mode.

4 Implementation of the Pocket Server

As stated above, the requirements of the Pocket Server are wide-ranging. However, present technology enables us to fulfill all of these requirements in a tiny single-board computer. As a concrete example, we present a sample implementation built on Raspberry Pi 3 Model B.[2] Figure 5 is a real image of this device, and Table 1 lists the specifications of Raspberry Pi 3 Model B (RP3). On this RP3, all requirements of the Pocket Server are realized with various open source software suites. The following are brief construction steps of the Pocket Server on RP3:

1. Format a new microSDHC and put it into the slot of the RP3.

Fig. 5 Raspberry Pi 3
Model B

Table 1 Specifications of
Raspberry Pi 3 Model
B. ARMv8-A (64/32-bit)

Architecture	ARMv8-A (64/32-bit)
SoC (system on chip)	Broadcom BCM2837
CPU	1.2 GHz 64-bit quad-core ARM cortex-A53
GPU	Broadcom VideoCore IV
RAM	1GB LPDDR2 (900 MHz)
On-board network	10/100Mbit/s Ethernet, 802.11n wireless, Bluetooth 4.1
On-board storage	microSDHC 16GB
Size	85.6 mm × 56.5 mm × 17 mm
Weight	45 g
Ports	HDMI, 4 × USB 2.0, Ethernet, and others

---

[2] Raspberry Pi 3 Model B is a product developed by the Raspberry Pi Foundation. https://www.raspberrypi.org/products/raspberry-pi-3-model-b/.

2. Install an operating system with NOOBS, which is an official operating system installer.[3] Here we choose the operation system, Raspbian, a Debian-based Linux operating system.[4]

3. Configure network interfaces appropriately.[5]

4. Install necessary software bundle, such as software development environments (C/C++, Java, Python, R, etc.), editors (Vim, Emacs, etc.), libraries, database applications (MySQL, postgreSQL, etc.), graphics-handling applications (ImageMagick, GIMP, etc.), applications for graph plotting (gnuplot, etc.), and text-processing applications (LaTeX, etc.), with the default package manager of Raspbian.[6]

5. Install other network handling applications (Cytoscape,[7] Gephi,[8] etc.).

The sample implementation in this section is mainly built by Python. For scientific computation required for network science, a lot of modules, such as NumPy, SciPy, matplotlib, NetworkX, pandas, and so on, are necessary. Though we can install these modules one by one with any of standard package managers of Python (e.g., by pip), here we use Berryconda,[9] which is an integrated Python distribution for scientific computing adapted for Raspbian based on the Conda package managing system, which includes all necessary modules.

The web application that plays a central role in presenting the contents of the course in the browser-oriented style is constructed on top of the Python module, Flask.[10] This module contains a tiny web server application of its own, which is sufficiently handy for present educational purposes. We do not need a gigantic server application such as Apache for this kind of job. Figure 6 is a screenshot of the entry page of the sample web application for this project. The web application can be made interactive on the server side implemented by Python, as well as on the client side using JavaScript. By using an appropriate database application, documents, graphics, and data necessary for the course are conveniently stored and retrieved from the web application.

Since the Pocket Server built on RP3 is a tiny but self-contained complete Linux server, the students can login to the server from their client machines and use the resources on the Pocket Server even in the server mode operation. The client mode operation is simply to use each Pocket Server as an independent Linux client PC.

---

[3] For NOOBS, see the web page at https://www.raspberrypi.org/downloads/noobs/.

[4] For Raspbian OS, see the web page at https://www.raspberrypi.org/downloads/raspbian/.

[5] Basic guides to configuring RP3 can be found in the web page at https://www.raspberrypi.org/documentation/configuration/.

[6] The Advanced Packaging Tool (APT) for Debian-based OS is also available in Raspbian. See the web page at https://www.raspberrypi.org/documentation/linux/software/apt.md.

[7] For Cytoscape, see the web page at http://www.cytoscape.org/.

[8] For Gephi, see the web page at https://gephi.org/.

[9] Berryconda is developed on GitHub at https://github.com/jjhelmus/berryconda.

[10] The main web page of Flask is at http://flask.pocoo.org/.

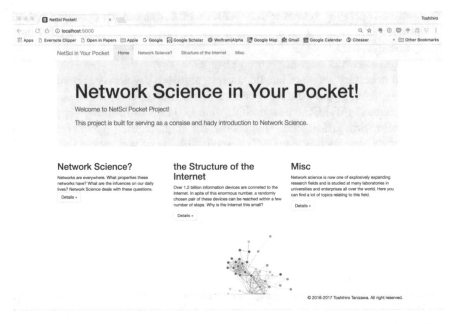

Fig. 6 Screenshot of the entry page of the web application "Network Science in Your Pocket!"

In a session of lectures, Pocket Servers are distributed to the students, and the students use those Pocket Servers as their client machines, in which a complete environment necessary for scientific computing in network science is provided.

5 Case Study of a Network Science Lecture with the Pocket Server

In this section, we briefly report on a network science lecture using the Pocket Server described in this chapter. The lecture was one of several open-house outreach activities at a technical college in Kochi, Japan, held in August 2016 for students from 10 to 15 years old, to promote their interest in science. The room for the lecture is equipped with 40 client personal computers with a projector and a large screen for the instructor. A private local area network connected to the Internet connection over a secure gateway is also available. Since we could not expect the participants to have prior knowledge of network science and sufficient skills for scientific computing, the lecture was planned in the browser-oriented style with three Pocket Servers of identical contents, operated in the server mode for dispersion of access requests from 40 client machines. Three fixed and different private IP addresses were assigned to these servers. Though the number of actual students who attended the course was about ten, the students were divided into three groups according to the number of the Pocket Servers and guided to connect to the server assigned to each group at the start of the lecture.

The contents of the lecture were presented basically by clicking the buttons of the web pages.[11] See Fig. 6 again as the entry web page for this lecture, though it should be noted that the actual web pages used in the lecture were localized versions in Japanese.

The aim of the lecture was to show the small-world property of scale-free networks due to the existence of hubs, by comparing the path length between two randomly selected nodes with that of two-dimensional regular lattices. The key question in this lecture was why the Internet, which consists of several billion of nodes, is so "small", in the sense that we feel almost no frustration in moving page to page by following the links on pages.

Next the instructor showed networks in the real world by showing several images such as the route map of the subway system in Tokyo, a road map of the city of London, a network of coauthorship of scientific papers, a network of the linkage connections between web pages on the Internet, the neuronal network of the human brain, Zachary's Karate Club network, a network formed by related English words, and so on.

After grasping the overall concept of networks, the students were led to lecture pages to learn the concepts of nodes, links, degrees, degree distribution, and path length between nodes. Interactive web pages written in JavaScript were also provided when they were considered to be effective. Figure 7 is an example of this kind

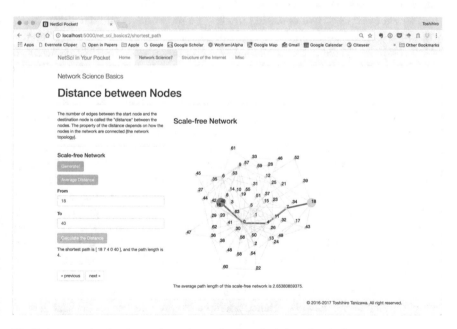

Fig. 7 An example of an interactive web page for the calculation of node distance

---

[11] The whole web application implemented by python-flask module can be downloaded from the author's GitHub page at https://github.com/toshitanizawa/NetSci-Pocket.

of interactive web page. According to the inputs from the students, the path length between the nodes are calculated and displayed dynamically. Finally, the students learned that the path length in scale-free networks does not increase proportionally to some power of the system size but logarithmically, which is the main reason for the smallness of the Internet.

Though all the students attending this lecture had no prior knowledge of network science, the survey after the lecture told us that they enjoyed learning various characteristics of networks, such as the universality of power-law behavior in the degree distribution or the small-world property. Above all, they were really surprised at the astonishingly small values of the path length in scale-free networks, knowing that they could not find two nodes whose path length was larger than 10, even in a huge network such as a Barabasi-Albert network of about 65,000 nodes.

6   Summary and Discussion

In this chapter, we describe building a small portable server that provides sufficient facilities for delivering lecture courses on network science, even if they are held off-campus. Thanks to current technology, it is possible to build such a server on a single-board computer so small that you can carry it in your pocket. A sample structure, which consists of two operating modes, the server mode and the client mode, was proposed, and an example of implementation of the server built on a single-board computer available on the market was also described. The effectiveness and capability of this server was actually tested at an open-house outreach activity at a technical college in Japan.

In this chapter the author reported a case study for using the Pocket Server in the browser mode. The browser mode is suitable for presenting ready-made contents that are well planned prior to the lectures. In contrast, the client mode is suitable for exploring or developing the style of the contents. Since the Pocket Server is a tiny but complete Linux machine provided with the necessary tools for scientific calculations and numerical simulations, students with sufficient skills can readily start a small research project on network science. The client mode operation of Pocket Servers could, therefore, be effective in an advanced or intensive program for network science targeted to highly motivated students in secondary schools. After instructions from the lecturer about the basic concepts of network science, the students would be introduced to selected topics of interest about network science and could conduct research on their own. You do not have to limit the use of the Pocket Server only to short lectures for outreach activities; an entire course on scientific computing including network science could be taught with Pocket Servers distributed to students.

The biggest challenge at present for the Pocket project is to provide sufficient lecture materials that are available to the teachers at any educational levels. Building a system for compiling the contents of outreach activities performed by members of NetSci community is crucial.

References

1. Erdös, P. and Rényi, A. (1959) On Random Graphs. I. Publicationes Mathematicae, 6. 290–297
2. Albert, R. and Barabási, A-L. (2002) Statistical mechanics of complex networks. Reviews of Modern Physics, 74:47–97.
3. Newman, M. (2003) The structure and function of complex networks. SIAM Review, 45:167–256, 2003.
4. Sayama, H., Cramer, C., Porter, M., Sheetz, L., and Uzzo, S. (2016) What are essential concepts about networks? Journal of Complex Networks, 4(3). 457–474.

Index

A

Academic programs
  academic minor, 61
  admission criteria and process, 75
  Center for Network Science (CNS), 89–91,
    93, 94
  Central European University (CEU),
    88–92, 94, 144
  curriculum
    applied mathematics, 4, 6, 9, 16, 19
    comprehensive exam, 92–94
    computational statistics, 75, 76, 78
    course materials, 26, 32
    course syllabi/schedules, 102, 104–106
    curricular sequences, 102
    data analytic courses, 76, 77, 79, 81, 83
    data collection, 102
    diffusion and influence, 111, 112, 114
    first year course work, 92
    linear algebra, 3, 9, 10, 15
    massive open online course (MOOC),
      178, 180–183
    Mathematical and Theoretical Physics
      (MTP) program, 10–12
    Mathematical Foundations of Computer
      Science (MFoCS), 7, 8, 10–12, 16
    mathematical graph theory, 189
    Mathematical Modeling and Scientific
      Computation (MMSC), 7, 11, 16
    methods of analysis, 102–104
    problem-driven, 88–91
    research advisor, 92
    research workshop and colloquium, 92
    teacher assistant (TA), 10
    workshop and colloquium, 92

degree requirements
  Academic Certificate, 23
  academic standards, 79
  credit hours, 79
  degree candidacy, 80
  dissertation advising, 79–80
  dissertation committee, 80
  dissertation defense, 75, 80
  doctoral program, 72–74, 79, 80, 82,
    84, 91
  qualifying examination, 80, 84
Dutch higher education, 46
  non-technical network literacy, 45–55
educators, 148, 150, 151
exam-based assessment, 11, 12, 14, 16
K-12
  classrooms, 151
  common core, 180
  curriculum, 142
  education, 142
  high school, 68, 160–162
  Network Science for the Next
    Generation (NetSci High), 11, 108,
    143–144, 149–155, 178, 180,
    190–193, 195–198
  Next Generation Science Standards
    (NGSS), 151, 180
  students, 142, 190
MFoCS students, 7, 8, 10–12, 16
miniproject-based assessment, 11, 12
miniproject group, 13
minor learning model
  learning experiences, 64
MMSC students, 7, 11, 16
Naval Postgraduate School (NPS), 23, 24

© Springer International Publishing AG, part of Springer Nature 2018
C. B. Cramer et al. (eds.), *Network Science In Education*,
https://doi.org/10.1007/978-3-319-77237-0

Academic programs (*cont.*)
    Academic Certificate, 23
    assignment, 24
    in-class conversations, 28
    in-class participation, 28–29
    learning and assessments, 28–40
    multilayer networks project, 27, 29–30
    team assignment, 30
    network scholars, 120
    Northeastern University, network science
        PhD program, 72, 73, 80, 84
    required core courses, 76–78, 81–83
    Oxford, University of, 4–6, 8, 11, 13–19
    PhD programs, 87–95
    candidates, 92–94
    colloquium, 92
    comprehensive exam, 80, 92–94
    course and related work, 93
    courses, 87, 91–93, 95
    curriculum, 72, 75, 81, 83, 91–93
    data collection and handling, 95
    doctoral-level education matching, 91
    economics, 88–94
    environmental science, 89–91
    experiences and outlook, 95
    first year work, research advisor, 92
    network science, 87–95
    research proposal, 92–93
    sociology, 88–90, 94, 95
    student recruitment, 94–95
    physical applied mathematics, 4, 16
    Queen Mary College, 95
    quizzes, 13, 14
    undergraduate programs
    curricula, 61–63
    disciplinary web, 61
    graph theory, 3, 4, 9, 16, 19
    Hilary Term, 8, 9, 11, 12
    Mathematical and Theoretical Physics
        (MTP) program, 10
    Mathematical Institute, 4
    Mathematical Modeling and Scientific
        Computation (MMSC), 7, 11, 16
    mathematics courses, 3–19
    Mathematics departments, 3, 4, 19
    Mathematical Foundations of Computer
        Science (MFoCS), 7, 8, 10–12, 16
    network science minor, 62–64
    United States Military Academy (USMA),
        59, 61–63, 69
    University of California, Los Angeles
        (UCLA), 9, 10, 13–15, 17–19
    Windesheim University, 45, 47, 49
Algorithms, 77, 79, 83

Ambiguity, 46, 48, 50
ANOVA analysis, 124, 126
    factorial analysis, 124, 126
Atlas of knowledge, 183

B
Brain Connectivity Toolbox, 17

C
Career goals, 160, 165
Code Book, 66
Cognitive science, 91
Complexity, 88, 92
Conferences
    Advances in Social Networks Analysis and
        Mining (ASONAM), 25
    International Conference on Complex
        Networks (CompleNet), 25, 145
    International School and Conference on
        Network Science (NetSci), 25, 87,
        89, 143–144, 149–152
        network science and education
        (NetSciEd), 150
    SIAM Workshop on Network Science, 25
    Sunbelt, 25
Connected: The Power of Six Degrees, 142
Connections: The Nature of Networks, 142,
    178–179
Cryptography, 66
Cyber-Enabled Discovery and Innovation, 144
Cyber Mission Forces, 66
Cyber operations, 63, 65–67
Cybersecurity and Cyberwar, 66
Cyber War, 66

D
Data analysis, 124–126
Data analytics and science, 66
Data collection, 102, 103
Data overlay design choices, 177
Data science, 3, 19, 88
Data set analysis, 33
Data visualization literacy, 169, 172,
    181, 182
Descriptives and correlations, 125
Descriptive statistics, 124–126
Dense networks, 160
Diffusion, 111, 112, 114
Dramatic improvements, 71
Dynamical processes, 76–78, 81
Dynamical systems, 108, 113

E

Economics and social psychology, 117
Ecosystems, 67
Ego-reciprocity, 125, 126, 130–133
    network position, 126, 130, 131
    OC and gender, 130, 133
Environmental science, 89–91
Epidemics, 112, 114
European Union Airline Data, 31
Evaluation, 146–148
Exam-based assessment, 11, 12, 14, 16
Experimental workshop elements, 50
Exploratorium, 142

F

Factorial ANOVA analysis, 124, 126
Flashcard app, 182, 183
Formal learning environments, 179–183
Forrester, Jay, 142
Foundational training, 73

G

Gender and network position, 121, 128–132
Geometry Playground, 142
Gestalt grouping, 184
Gestalt laws, 174
Global Maritime Transportation
    network, 34
Graph aesthetic principles, 176
Graph limits, 89
Graph theory, 3, 4, 9, 16, 19, 77, 79, 88, 89,
    91, 93, 141, 142, 153
Guided discovery, 23

H

Hand-drawn network diagram, 50, 51, 53
Higher income group, 131
Higher professional education, 46–47, 53
Homophily, 121
Human capital, 117, 119, 126

I

Individual salary, 119
Informal learning environments, 178–179
Information Visualization MOOC, 178,
    180–183
Instruments, 122
Integration course, 64
Intellectual capital, 118–119, 134
Interactive teaching style, 28

Interdisciplinary, 59, 60, 63–65, 67, 68, 72–74,
    83, 84
Interdisciplinary analysis, 60
Interdisciplinary Contest in Modeling (ICM),
    67, 68
Internet of Things, 66
IVMOOC course, 182, 183
IVMOOC flashcard app, 182, 183

K

Kolmogorov-Smirnov, 147

L

LAWR ego-reciprocity, 125, 126, 130–133
LAWR network, 125–128, 134
LARW Network of Principals, 126
Leadership advice (LA) network,
    123–125
Linear algebra, 3, 9, 10, 15
Local area network (LAN), 191, 192, 196

M

MapStats Curriculum, 142
Massive open online course (MOOC), 178,
    180–183
Mathematical graph theory, 189
McKinsey Global Institute (MGI) study, 72
Mega Mathematics Project, 142
Metaphoric Literacy, 173–174
MIT Media Lab, 143
Modeling, 4, 5, 8, 9, 12, 15–19
Modularity maximization method, 104, 105, 107
Molloy-Reed configuration model, 24
Multidisciplinary, 89
Multilayer network, 27, 29, 31
Multilayer Networks Project, 27–30

N

National Institutes of Health, 72
National Research Council, 72
National Research Council Network Science
    Committee, 60, 61
National Science Foundation, 72
Naval Postgraduate School (NPS), 23, 24
NetSci community efforts, 150
Network analysis methods, 101–102
Network and information science, 60–61
Network data properties, 173, 174, 177
Network diagram, 49–55
Network economics, 78–79

Network literacy, 150, 151, 153–155, 190
    essential concepts and ideas, 54, 150, 151,
        153–155
    workshops, 49, 50, 53, 54
Network modeling, 72, 75, 76, 78, 143, 144
    applications, 190
    Barabasi-Albert network, 24, 26, 198
    Bayesian, 78
    clustering, 48, 52
    communities, 101, 105, 107–109, 113–115
    complex networks, 23, 25, 26, 32
    edge weights, 104–106, 109, 110
    Erdos Renyi models, 108, 112
    exponential random graph model
        (ERGM), 78
    directed, 101–104, 115
    features, 189
    fundamentals, 64–65
    multilayer, 27, 29, 31
    network dynamics, 3, 5, 6, 50, 53, 55
    network structure, 101, 102, 107–109,
        111–115
    power-law, 198
    random networks, 24, 26, 105, 107–109,
        111–115
    scale-free networks, 111–113
    Seven Bridges of Konigsberg, 23, 189
    small-worlds, 24, 26, 105, 108, 111–113,
        115, 189, 197, 198
    spanning tree, 104, 109–115
    Watts-Strogatz, 24, 26
Network Profile Summary, 24, 25, 27, 28, 33–41
Network representation, 107–109, 112–114
Network scholars, 120
Network science, 141–151
Network science community, 101, 115
Network Science Data I, 76, 77, 81, 83
Network Science for the Next Generation,
        144–146, 149
Network Science Institute, 73
Network Visualization Education, 181–182
    adjacency list, 170
    data visualization, 169, 172, 174
    formal learning environments, 179–183
    Ghoniem, Fekete and Castagliola, 177
    informal learning environments, 178–179
    matrix visualization, 170
    node-link diagram, 170, 171, 173, 175, 177
    Purchase, Cohen, and James, 177
    representational literacy, 172–173
    teaching network visualization literacy,
        177–183
    topological Literacy, 174–177
    visualizations, 48, 52, 55, 169–185

Network visualization literacy (NVL),
        169–185
Newman, Mark, 5, 17, 26
New York Hall of Science, 142, 178, 180

O
Offensive cyber operations, 66
Organizational commitment (OC), 117, 118,
        120, 122–135
Outreach education activities, 190
Outstanding-category solution, 67, 68

P
Phase transition, 112, 114
Pocket project, 189–198
    client mode, 192, 193, 195, 198
    Pocket Server, 190–198
    Raspberry Pi, 194
    server mode, 192, 193, 195, 196, 198
Political science, 89–91
Probability, 3, 4, 9, 10, 15, 16, 19
Proceedings of the National Academy of
        Sciences, 8, 13
Professional development, 151
Program Evaluation Survey, 82–83
Programming, 71–85
Project-based learning, 24

Q
Quadratic assignment procedure (QAP), 124, 125

R
Real-world complexity, 45, 46
Real-world problems, 45–48, 141, 143, 144
Reciprocal ties, 118, 130, 134, 135
Research Project, 24, 28–33, 41

S
Salary and social network position, 119
Sample and context, 121–122
School principals, 118–122
Setting goals, 161–162
Social capital, 117–119, 159–166
    bonding, 159
    bridging, 159, 160
    capital assets, 117, 119, 135
    ideal board of directors, 162, 166
    intention, setting, 164
    real board of directors, 163–164

Social networks, 117–124, 135, 159–161, 164
    actor, 118, 119, 125
    centrality, 103, 105, 107–109, 112–114
        betweenness, 175
        closeness, 125
        incloseness, 125
    Karate Club network, 197
    network position, 117–121
    and social capital, 159–161, 164
    social network analysis (SNA), 141, 147, 152
    strong ties, 160, 161, 163
    two-plex ties, 124
    weak ties, 160, 161, 163
Social sciences, 60, 66, 67
Sociology, 88–90, 94, 95
Software
    C, 190, 195
    C++, 190, 195
    Gephi, 24, 26, 27, 31, 145
    igraph, 26
    Java, 190, 195
    LaTeX, 24, 27
    Mathematica, 104
    MATLAB, 17
    NetworkX, 26, 145, 180
    Organizational Risk Analysis
        (ORA), 64
    Python, 77, 145, 148, 149, 190, 195, 197
    R, 24, 26, 30

    Sci2, 183
    UCINET 6.0, 125
Statistical Mechanics, 9, 16
Statistical Physics of Complex Networks,
    78, 83
STEM Workforce, 141
Stochastics, 108, 113
Summative evaluation, 147
Synthetic network models, 26
Systems thinking, 50

T
Title 1, 145
Two-plex network, 123–126
Two-step approach, 51

U
UMAP Journal, 67, 68
Underrepresented, 143, 145, 149

V
Visual Insights textbook, 183

W
Work recognition (WR) network, 123–125

Printed in the United States
By Bookmasters